LAW ENFORCEMENT AGENCIES

CRIME LAB

LAW ENFORCEMENT AGENCIES

Bomb Squad

Border Patrol

Crime Lab

Drug Enforcement Administration

Federal Bureau of Investigation

Interpol

Los Angeles Police Department

New York Police Department

The Secret Service

SWAT Teams

The Texas Rangers

U.S. Marshals

LAW ENFORCEMENT AGENCIES

CRIME LAB

Colin Evans

CRIME LAB

Copyright © 2011 by Infobase Learning

All rights reserved. No part of this book may be reproduced or utilized in any form or by any means, electronic or mechanical, including photocopying, recording, or by any information storage or retrieval systems, without permission in writing from the publisher. For information contact:

Chelsea House
An imprint of Infobase Learning
132 West 31st Street
New York NY 10001

Library of Congress Cataloging-in-Publication Data

Evans, Colin, 1948-
 Crime lab / Colin Evans.
 p. cm. — (Law enforcement agencies)
 Includes bibliographical references and index.
 ISBN-13: 978-1-60413-612-8 (hardcover : alk. paper)
 ISBN-10: 1-60413-612-X (hardcover : alk. paper) 1. Crime laboratories
 2. Evidence, Criminal. 3. Criminal investigation. I. Title. II. Series.
 HV8073.E923 2011   363.25—dc22
             2010043271

Chelsea House books are available at special discounts when purchased in bulk quantities for businesses, associations, institutions, or sales promotions. Please call our Special Sales Department in New York at (212) 967-8800 or (800) 322-8755.

You can find Chelsea House on the World Wide Web at
http://www.infobasepublishing.com

Text design and composition by Erika K. Arroyo
Cover design by Keith Trego
Cover printed by Bang Printing, Brainerd, Minn.
Book printed and bound by Bang Printing, Brainerd, Minn.
Date printed: March 2011

Printed in the United States of America

10 9 8 7 6 5 4 3 2 1

This book is printed on acid-free paper.

All links and Web addresses were checked and verified to be correct at the time of publication. Because of the dynamic nature of the Web, some addresses and links may have changed since publication and may no longer be valid.

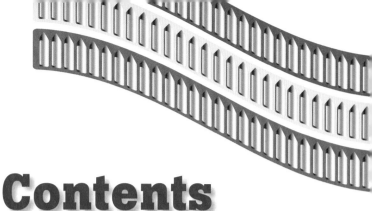

Contents

Introduction	7
1 The Early Days	17
2 The Birth of Ballistics	27
3 You Are What You Eat	38
4 Atomic Evidence	49
5 A Disputed Document	60
6 Bags of Evidence	71
7 Time of Death	82
8 A Valentine's Day Massacre	94
9 Prints and Pixels	106
Chronology	118
Endnotes	122
Bibliography	124
Further Resources	126
Index	128
About the Author	134

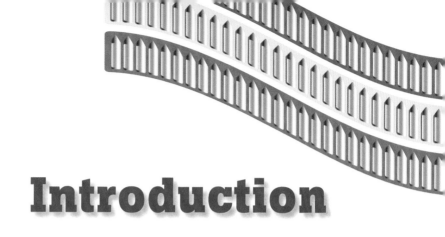

Introduction

In the world of modern law enforcement, the role of the crime laboratory has become crucial, especially if a case goes to trial. Prosecutors recognize that their best chance of impressing a jury—and thereby securing a conviction—rests in being able to introduce some kind of science-based evidence that links the accused to the crime. Juries nowadays expect it. Indeed, so widespread has this perception become that the courts even have a name for it: the "*CSI* Effect." Prosecutors and defense attorneys alike realize that most jury members will at one time or another have seen an episode of the extraordinarily popular TV franchise, and—if not consciously, then certainly subconsciously—many on that jury will expect the same level of scientific wizardry as that displayed weekly by Grissom, Caine, and the rest of the gang. It's wholly unrealistic, of course, and most in the law enforcement community loathe this perceived hijacking of the legal process, but this is the reality, the expectation. In the eyes of the general public, from whom all jury members are chosen, the crime lab is expected to solve crimes. How often, after a trial in which a defendant has been acquitted, has a jury said, "The prosecution had nothing, no forensics . . .?"

This dependence did not arise overnight. It took a long time for the courts, and more importantly juries, to accept scientific evidence. In Europe during the 19th century, a string of disasters involving expert witnesses cast a pall over any type of scientific testimony. This led to many miscarriages of justice. Poisoners walked free and, in one notorious English case, an American woman, Florence Maybrick—convicted on the flimsiest toxicological evidence of murdering her husband—was rescued from the hangman's noose only at the very last moment, after doubts emerged about the quality of some of the expert testimony. (She still served 15 years before being released.)

8 CRIME LAB

A crime lab technician examines evidence following a shooting in Philadelphia that left one young man dead and another critically wounded. *(AP Photo/Joseph Kaczmarek)*

In Mrs. Maybrick's homeland, things were hardly any better. American forensic science was slow out of the blocks and all too often degenerated into long-winded courtroom battles between overblown egos. Muddled medical testimony ended up causing more problems than it solved.

The big transformation came with the dawn of the 20th century. Before this, there had been breakthroughs in macro-detection—the first reliable toxicology test and recognition of fingerprinting as an identification tool—but it was the huge technological leaps of the new century that made the modern crime lab possible. The electron scanning microscope, chromatography, the electrostatic detection apparatus, DNA typing, neutron activation analysis—these are just a few of the weapons that the crime scientist can call upon, and all have been invented in the last 100 years.

Modern crime scenes resemble something from the pages of science fiction. Ghostly characters in white Tyvek suits and face masks move stealthily about, bagging and tagging those scraps of evidence that appear almost invisible to the untrained observer. These are the investigators we associate with the miracle of forensic science. But in most cases it will be the crime lab that makes the final call. It is their job to analyze any evidence recovered from the crime scene, and they get to decide its real evidential value. To achieve this they employ a bewildering array of equipment that, once again, smacks of space-age technology. For instance, at the Royal Armaments Research and Development Establishment at Fort Halstead in Kent, England—arguably the most sophisticated crime lab in the world—the test benches have zero static electricity and are equipped to the same standard as NASA. Very few authorities can afford this level of investment, but it does give an idea of the lengths to which governments are prepared to go in order to catch the bad guys.

Nowadays, because of the prohibitive cost of cutting-edge forensic science and the rapid pace of development, many crime labs have chosen to specialize in one area. This is especially true of DNA analysis. Not so long ago, large amounts of high-quality human samples were necessary to produce a DNA profile. By the early 2000s the quantity required for conventional DNA testing had shrunk to 50 to 100 human cells (roughly equivalent to the smallest bloodstain visible to the naked eye). Now, thanks to what is known as low copy number analysis (LCN), a profile can be compiled from a mere five human cells. To put this in perspective, this is the amount of skin cells that might be left on a smudged fingerprint. For some, this is a step too far. There are claims that a person might shake hands with an individual and then that person could go and commit a crime, leaving traces of the innocent person's DNA at the scene. While such claims have yet to be substantiated, they do raise legitimate concerns. The Federal Bureau of Investigation (FBI), for one, is unimpressed by LCN analysis and, at the time of this writing, does not use the technique. The FBI's reticence highlights an ongoing problem with forensic science: The advances are coming at such a pace that the legal system is struggling to keep up.

HOW CRIME LABS FUNCTION

Crime labs exist to answer two fundamental questions:

1. What is the sample being tested? Is it blood, for instance, or merely a splash of wood dye?
2. Can this sample be definitely linked to a suspect?

The following is a list of the most common forms of evidence examined by crime labs:

- ★ Anthropological finds (body parts)
- ★ Arson/fire-related residue
- ★ Biological samples (bodily fluids)
- ★ Blood
- ★ Documents
- ★ Drugs and related paraphernalia
- ★ Drunk driving samples, usually blood and urine
- ★ Explosives and explosive residue
- ★ Firearms
- ★ Glass
- ★ Glove-prints
- ★ Hair/fibers/dust
- ★ Images that require digital enhancement
- ★ Impressions such as shoe prints, tire prints
- ★ Paint
- ★ Poisons
- ★ Shoes/footwear
- ★ Soil samples

Introduction 11

The crime lab has not been around a long time, but in its relatively short life it has proved itself invaluable to prosecutor and defense attorney alike. Without it, the modern justice landscape would look very different indeed. Skeptics who constantly bemoan what they perceive to be "junk science" should consider the following figures, published by the FBI: In America in 2007, someone was murdered every 31 minutes; each hour saw 12 women sexually assaulted; and 1 violent crime

★ Textiles
★ Tool marks
★ Vehicles
★ Weapons other than firearms, most commonly knives

Identifying the sample is only half the battle; assessing its evidential value is when the real struggle begins. If a trace of blood, say, has been found at a crime scene and it does not belong to the victim, but it can be proved to come from a suspect, then this is strong evidence to suggest that person's presence at the crime scene. This is quite straightforward. Problems start to arise when the focus shifts to the area of evidence interpretation. Give two scientists the same set of results and it is entirely conceivable that two opposite conclusions will be reached. If, for example, a fine mist of blood spray is found on the jacket of a person accused of battering someone to death, both scientists may agree that it is blood from the victim, but one may believe that the stain has been transferred during the murderous attack. The other scientist disagrees. He argues that the blood particles were expelled on the breath of the dying victim, and had been transferred innocently to the suspect as he bent over to tend to the victim's injuries. (For the record, this scenario actually happened. A jury listened to the evidence and rendered a guilty verdict, only to have it overturned on appeal.) This gives some idea of the difficulties facing crime labs and courtrooms.

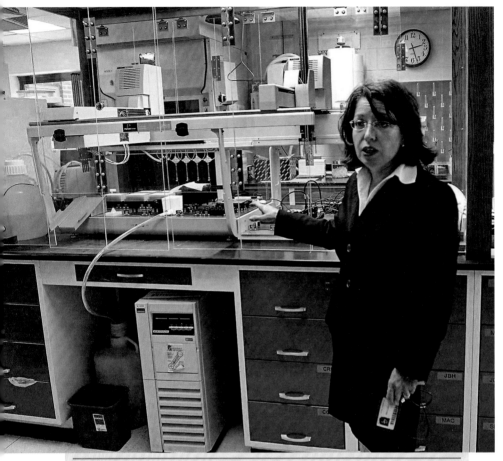

Soraya McClung, director of the West Virginia State Police Crime Lab, shows off the DNA lab at the police headquarters in Charleston, West Virginia, in May 2010. *(AP Photo/*The Charleston Gazette, *Chris Dorst)*

occurred every 22.4 seconds.[1] These are horrifying numbers. Just imagine how much more difficult crime investigation would be without the crime lab.

Crime Lab explores the work performed by crime lab personnel and its importance in the context of a criminal investigation.

Chapter 1, "The Early Days," surveys the origins of the crime laboratory and the need for a more scientific approach to crime solving.

Chapter 2, "The Birth of Ballistics," discusses how scientists began matching bullets to guns, culminating in the invention of the comparison microscope.

IT'S IN THE BLOOD

Of all the types of trace evidence that a crime lab is asked to analyze, few are more informative or productive than blood. It can place someone at a crime scene, and it can, through bloodstain pattern analysis, sometimes paint a picture of how the crime unfolded. Most forms of violence will result in bloodshed. As the average adult has about 10 pints of blood running through his or her system, it is little wonder that, when punctured arterially, the human circulatory system has a tendency to spray its contents like a water fountain.

Although scientists in the 19th century recognized that blood consists of various types, it was not until 1901 that Viennese immunologist Karl Landsteiner standardized the modern ABO system. By separating the serum from red blood cells in a centrifuge, then adding red blood cells from different people to the serum, he found that two distinctly different reactions occurred. In some cases the serum seemed to attract the red blood cells, but in others it repelled them. One lot of cells agglutinated—or clumped together—and others did not.

Landsteiner labeled these two blood types A and B, only to soon realize that there was a third type that did not react like either A or B, but which showed characteristics of both. Although he called this group C, it soon became more popularly known as O. One year later, an assistant discovered yet another serum type that did not agglutinate with either A or B. This was labeled AB. Allowing for some racial and geographical variation, blood types are normally distributed in the population as follows:

Group O: 43–45 percent
Group A: 40–42 percent
Group B: 10–12 percent
Group AB: 3–5 percent

In the mid-1920s Landsteiner discovered another grouping system as a result of injecting rabbits with human blood. This

(continues)

14 CRIME LAB

(continued)

produced M and N groups; a person could be either M or N, or a combination of the two, MN. Modern science has continued Landsteiner's work to the point where there are now 257 antigens (a substance that prompts the generation of antibodies and can cause an immune response), and no fewer than 23 blood group systems based on association with these antigens. In 1930 Landsteiner was awarded the Nobel Prize for Medicine for his groundbreaking work in this field.

Although initially conducted to eliminate the dangers of indiscriminate blood transfusion, Landsteiner's research, and that of his followers, has had an enormous bearing on the course of medical jurisprudence to the present day.

When confronted by a suspect stain, the serologist (a medical scientist specializing in blood serum) must ask four distinct questions:

1. Is the sample blood?
2. Is the sample animal blood?
3. If animal blood, from what species?
4. If human blood, what type?

The first question is answered by employing color or crystalline tests. Benzidine was popular until its toxicity was discovered. Today scientists use a chemical called phenolphthalein. When it comes in contact with hemoglobin (and sometimes other organic matter), phenolphthalein turns the sample a bright pink color. In a similar test, using orthotolodine, the sample turns bright blue. Even bloodstains invisible to the naked eye cannot escape detection. For this there is luminol, a chemical that when sprayed on walls, floors, and furniture in

Chapter 3, "You Are What You Eat," demonstrates the efforts made by New York City to catch up with the major European cities in its adoption of a dedicated crime laboratory.

the dark reveals a slight phosphorescent light where bloodstains (and certain other stains) are present.

To answer questions 2 and 3, forensic scientists use antiserum or gel tests. The standard test—called the precipitin test—originally involved the use of rabbit serum. Nowadays it has been greatly simplified, and involves passing an electric current through a suspected blood sample that has been treated with gelatin. If a precipitin line forms at a specified point, this indicates that the sample is human blood.

The answer to question 4 depends on the sample being both an adequate volume and of sufficient quality. If this is so, direct typing using the ABO system is done. Indirect typing would have to be done on severely dried stains. The most common indirect typing technique is the absorption-elution test, which is performed by adding compatible antiserum antibodies to a sample, heating the sample to break the antibody-antigen bonds, and then adding known red cells from standard blood groups to see what coagulates.

Blood at the crime scene can be in the form of pools, drops, smears, or crusts. Of these, wet blood has the most evidentiary value because more tests can be run. For example, alcohol and drug content can be determined from wet blood only. Blood begins to dry after three to five minutes of exposure to air, and as it dries, it turns brown or black.

Besides blood, crime lab serologists will analyze semen, saliva, and all other bodily fluids. Such diversity inevitably leads them into the associated discipline of DNA typing, and occasionally bloodstain pattern analysis as well, though the latter has tended to become more of an independent specialty.

Chapter 4, "Atomic Evidence," shows how scientists have harbored nuclear energy to develop one of the most sophisticated forms of chemical identification yet devised.

Chapter 5, "A Disputed Document," explores the multimillion-dollar world of art forgery though a baffling case that continues to spark controversy.

Chapter 6, "Bags of Evidence," provides an insight into how extraordinary laboratory analysis of materials and machinery managed to trap a pitiless killer.

Chapter 7, "Time of Death," reveals how not all crime labs work under optimum standards of hygiene and cleanliness. In Tennessee there exists, arguably, the most bizarre crime lab in the world, but one whose work is revolutionizing knowledge of what happens to human bodies after they die.

Chapter 8, "A Valentine's Day Massacre," delves into the ways that arson investigators in Pittsburgh investigated the deaths of three firefighters who fell victim to a callous insurance scam.

Chapter 9, "Prints and Pixels," shows how the crime lab is always coming up with ever more inventive ways of trapping criminals. Fingerprint evidence is one of the oldest branches of forensic science, but investigators in Washington State managed to add a new weapon to their arsenal through the use of computer enhancement.

The Early Days

It was at the end of the 18th century that scientists in Europe first began to explore the possibilities of using the laboratory in the war on crime. This arose primarily out of a desire to stop poisonings. Unbelievable as it might seem today, for most of recorded history it was impossible to detect the presence of poison in the human body. This immunity gave the poisoner a license to kill. As a result, countless numbers of people were murdered without anyone being the wiser. Most victims of poisoning succumbed to the effects of arsenic. There was a good reason for this. Odorless and tasteless, arsenic is also a lethal mimic. In the Middle Ages if someone spent a few days in bed rolling in agony and retching violently before expiring, chances are that death would have been attributed to some gastric disorder or maybe drinking contaminated water. Few would have suspected poison. No one, of course, knows just how many murders went unrecorded because of this ignorance, but the great royal families of Europe were certainly aware of arsenic's efficacy. Catherine de Medici, who married Henry II of France in 1533, was credited with an in-depth knowledge of all types of poison, while some families were even rumored to keep professional poisoners on the payroll to eliminate rivals. Not for nothing did arsenic become known as "inheritance powder" for the way in which it shaped the direction of dynasties and the contents of their bank accounts.

The fight against the poisoner began in 1787 when a German chemist, Johann Metzger, discovered that if substances containing arsenic were heated and a cold plate held over the vapors, a white layer of arsenious oxide would form on the plate. While this "arsenic mirror" could establish whether food had been laced with arsenic, it could not tell if a body had already absorbed arsenic.

The solution to this problem was provided in 1806 by Dr. Valentine Rose of the Berlin Medical Faculty. He took the corpse's stomach with its contents, cut them up, and boiled them into a kind of stew. After filtering the stew to remove any remaining flesh, he then treated the liquid with nitric acid. This had the effect of converting any arsenic present into arsenic acid, which could then be subjected to Metzger's "mirror" in the usual way.

But by far the greatest toxicological leap forward came in 1836, when James Marsh, a London chemist, invented a means of detecting even the smallest quantity of arsenic. It was similar to Metzger's

A toxicologist conducts research on arsenic, using the apparatus devised by James Marsh, in Paris in 1928. *(Roger Viollet/Getty Images)*

THE FIRST GREAT CRIME CHEMIST

James Marsh was born in London in 1794. He studied chemistry and worked alongside the celebrated inventor Michael Faraday at the Royal Military Academy, Woolwich. In November 1833 Marsh was asked to analyze some coffee that a George Bodle of Plumstead had drunk just before his death. Marsh found arsenic present—as he testified to the inquest jury, which returned a verdict of willful murder against Bodle's grandson, John Bodle. Despite Marsh's testimony, when Bodle came to trial he was acquitted. (Some months later he confessed to the crime.)

This fiasco infuriated Marsh, who turned his attention toward the problem of detecting arsenic. He was determined to improve the process. By 1836 he had developed a method of detecting arsenic in quantities as minute as 1/50th of a milligram. His peers were astonished. For this extremely sensitive test he received the gold medal of the Society of Arts that year. He continued improving the test in the ensuing years, and in 1840 it came to widespread public notice during the trial of Marie Lafarge in Tulle, France, for the murder of her husband. The original forensic examination found no trace of arsenic in his body, but the court asked for a new test. The body was exhumed and was found, using Marsh's test, to contain arsenic. This evidence convicted Lafarge, who was sentenced to life imprisonment. The case caused a sensation on both sides of the English Channel and ensured that Marsh's name became a familiar one to the public.

Marsh, meantime, had gone back to the laboratory where, in 1837, he invented a percussion cap for naval guns for which he won the silver medal of the Society of Arts and was rewarded financially by the Admiralty. But his final days were difficult, and when he died on June 21, 1846, his wife was left penniless. In recognition of Marsh's achievements—including his invention of the arsenic detection test—the government later awarded his widow a pension.

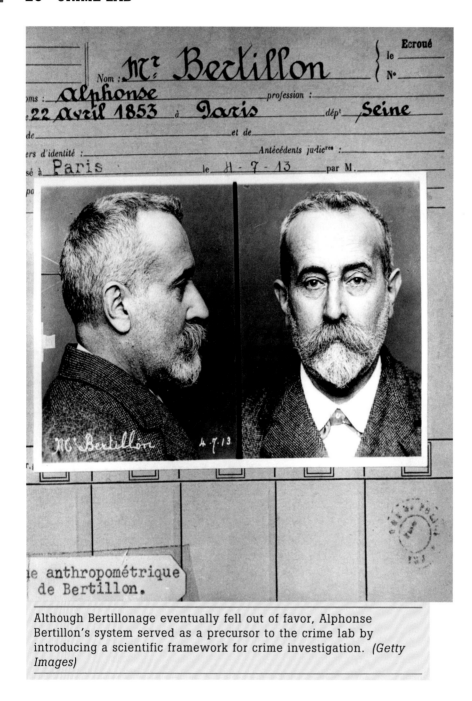

Although Bertillonage eventually fell out of favor, Alphonse Bertillon's system served as a precursor to the crime lab by introducing a scientific framework for crime investigation. *(Getty Images)*

method, but instead of allowing the vapors to rise up to the cold metal plate—with most of the gases escaping into thin air—the whole process took place in a sealed U-shaped tube that only allowed the vapors to exit via a small nozzle. The suspect material was dropped onto a zinc plate covered with dilute sulfuric acid to produce hydrogen. Any arsine gas was then heated as it passed along a glass tube, condensing when it reached a cold part of the tube to form the "arsenic mirror." In a refined form, the Marsh test is still in use today.

Marsh's invention provided the catalyst for an explosion in science-based crime detection, most of which took place in France. Leading the charge was Alexandre Lacassagne, a former army physician who, in 1878, published a forensic manual called *Précis de Médicine Légale,* the outstanding success of which led to Lacassagne being offered the newly established chair in Forensic Medicine at the University of Lyon. Lacassagne was a brilliant man with a broad body of knowledge: his military experience gave him an unrivaled knowledge of gunshot wounds; he knew more about postmortem changes in the human body than anyone else alive; and he was a pioneer in the study of crime scene blood spatter analysis. And yet, oddly enough for someone of his undoubted brilliance, Lacassagne featured in very few criminal cases. This was due to the intransigence of the French police, who were reluctant to call on the services of a mere academic to help solve crimes. As a result, Lacassagne's laboratory at Lyon investigated few criminal cases and was geared more toward teaching.

It would take another French forensic pioneer, Alphonse Bertillon, to break down the state-sponsored inertia in France. Despite Bertillon's influence, there was still no cohesive attempt to centralize the study of forensic science. It remained the reserve of university academics and wealthy gentleman scientists. As early as 1849, a Massachusetts dentist had used dental records to identify the remains of missing Bostonian George Parkman—evidence that doomed a Harvard chemistry professor named Joseph Webster to the gallows—but since that time American interest in forensic science had tapered off disappointingly. Not even the single most important discovery in the history of crime fighting—that no two humans share the same fingerprints—could light a fire under the authorities. Decades of infighting would follow before the reliability of fingerprinting was generally accepted, and by that time it was the early 20th century.

(Continues on page 24)

THE MAN WITH THE CALIPERS

Although he did not operate a crime lab *per se,* Alphonse Bertillon was one of the first to bring a thoroughly scientific approach to the business of scientific crime detection. His father was president of the Paris Anthropological Society, while his brother, Jacques, became an eminent statistician, and it was through a combination of these two disciplines that Alphonse made his name. An undistinguished scholar—he was expelled from the Imperial Lycée of Versailles—Bertillon fared no better in his first job (he was dismissed from a Paris bank) and only parental string-pulling secured him a post as assistant clerk in the records office at the Prefecture of Police.

His duties included itemizing the particulars of arrested felons and entering them on a record card. The work took him to the notorious La Santé Prison, which housed some of France's most dangerous criminals. He became obsessed with the notion that no two humans share exactly the same physical characteristics. From this belief, he evolved a system—which he dubbed "anthropometry"—of recording bodily measurements.

He first categorized a person in one of three main head-size types. Next, the subject was subdivided according to the dimensions of the bony parts of the body, in an 11-step process. Special attention was paid to distinguishing marks such as moles, tattoos, and scars, and especially eye color. The hundreds of calculations that resulted provided the ammunition for Bertillon's boast that the chances of two people sharing identical physical characteristics were less than one in 4 million. Finally, he refined traditional photo mug shots to include two views—frontal and profile—a system still in use throughout the world.

In 1879 he published his findings. His employers were not impressed. Another three years of refinement were demanded before the Paris police condescended to give what became known as *Bertillonage* an extended trial. Given the absence of any previous means of coherent criminal identification, such official resistance is hard to fathom, but gradually Bertillon prevailed. In its first three years of use, anthropometry was responsible for more than 800 arrests, as criminals found that they could no longer hide their real identities from Bertillon's calipers and compasses.

One great advantage of Bertillonage was the ease with which a criminal's details could be telegraphed to various parts of the country. Such speed led to the cracking of many cases that would otherwise have gone unsolved.

Bertillon's star shone brilliantly through the 1880s, as police forces around the world embraced his revolutionary method. Appointed director of the Judicial Identification Service, his finest hour came in 1892 when he identified the notorious anarchist called "Ravachol" as mass murderer François Claudius Koenigstein. For this, Bertillon was awarded the Legion of Honor.

Such triumphs only fueled Bertillon's belief in his own forensic omnipotence. He began to meddle disastrously in areas outside his specialized field. It was his unhesitating identification of Alfred Dreyfus, a French army officer accused of passing secrets to Germany, as the author of a traitorous letter, that led to the wholly innocent Dreyfus spending 12 years on Devil's Island before the real culprit was named. Other blunders followed.

It was the introduction of fingerprinting that really finished off Bertillon. He fought stubbornly and hard against the new identification tool, convinced it was inferior to his anthropometry. But the rest of the world did not agree, and Bertillon was swept aside. Marginalized and increasingly ignored, he faded into embittered obscurity until his death at age 60 in 1914. He is buried alongside the likes of Chopin, Sarah Bernhardt, and Jim Morrison in Paris's celebrated Père Lachaise cemetery.

(Continued from page 21)

It was at this time that changes in attitude began to be noticed. As mentioned earlier, this was an era of great technological innovation. Also, scientists were now gaining insights into human biology that put a greater strain on the laboratory's resources. Karl Landsteiner, in Austria, had unlocked the secrets of the blood grouping in humans. Across the border, the German biologist Paul Uhlenhuth devised a test to distinguish human blood from that of other animals. There was now so much science around that law enforcement agencies could no longer afford to ignore it.

Ironically, to find the true pioneer of the dedicated crime laboratory one must go back to Lyon and to the classroom of Alexandre Lacassagne. One of Lacassagne's brightest pupils was a young man named Edmond Locard, and over the course of several years Locard rose to the rank of Lacassagne's assistant. In 1910 Locard resigned the post to establish what became known as the Laboratoire Intérregional de Police Technique in Lyon.

The title sounded impressive, but the reality was more modest. In fact, Locard's workplace was a simple two-room operation on the second floor of the Lyon courthouse. The only two items of equipment he had were a microscope and a spectroscope. Locard's laboratory might have been humble, but it has earned its place in history as the world's first dedicated crime lab. Every state-of-the-art medico-legal facility around the globe can trace its pedigree back to the Lyon courthouse.

Locard was convinced that the future of crime detection lay in rigorous and scientific analysis of evidence. And over the course of time he evolved a theory that would form the bedrock of forensic science. It became known as Locard's Exchange Principle, and put in its simplest form it states: *Every contact leaves a trace.* At its core was Locard's belief that whenever two human beings come into physical contact, no matter how briefly, something—a fiber or hair, for example—from one is transferred to the other. The axiom is as true today as it ever was.

It did not take long for the notion of the stand-alone crime investigation facility to spread worldwide. In the beginning, many were "one-man" operations, often tucked away on university campuses. A typical example was founded in the 1920s by an extraordinary scientist named

Edmond Locard, French doctor and criminal jurist, is best known for his Exchange Principle, which states that every contact leaves a trace. *(Collection Roger-Viollet/The Image Works)*

Edward Oscar Heinrich, professor of chemistry at the University of California at Berkley. Heinrich solved scores of crimes single-handedly and came up with so many innovations that he was dubbed the "Edison of crime detection."[1] But he was almost the last of a dying breed. The death knell of the one-man crime laboratory was rung in 1932, when the United States Bureau of Investigation (forerunner of the FBI) established its Technical Laboratory. Although initially operated strictly as a research facility, the Technical Laboratory soon began compiling extensive reference collections of guns, watermarks, typefaces, and automobile tire designs, while its lavish federal funding allowed for the purchase of specialized microscopes and the latest technological advances.

Where the FBI led, other agencies followed. Crime labs began springing up across the United States, most under the control of state authorities. In recent years, however, there has been a resurgence in the number of private facilities, as companies have recognized the profit-making potential of forensic science and governments have realized the advantages of specialization.

The Birth of Ballistics

Although the term *ballistics* actually refers to the science that deals with the motion of projectiles and the conditions affecting that motion, in criminal investigation it has come to mean the specific study of firearms and bullets. Generally, this involves matching a bullet to a particular gun. What forensic scientists nowadays take for granted is actually quite a recent innovation, although the origins of firearms identification can be traced back to the late 15th century. This was when German gunmakers employed by the Holy Roman Emperor Maximilian I discovered that by etching a spiral groove along the length of the barrel of a firearm, they could dramatically improve that weapon's accuracy. After being smooth-bored, the gun barrel blank is reamed to specification diameter, and then rifled. The raised parts of these rifling marks are called lands, while the valleys are referred to as grooves. As the bullet hurtles through the barrel, the rifling marks impart spin, and it is this rotation that helps the bullet hold its trajectory, enabling it to travel straighter for longer. This was bad news for the person on the receiving end, and great news for the modern crime lab. For what Maximilian's gunsmiths could not have imagined was that rifling has another benefit: It makes bullets identifiable.

The identification of bullets is made possible by an unavoidable manufacturing defect. Each time a barrel is rifled, the machine tool that gouges through the metal wears ever so slightly. As a result, when the next barrel is being rifled, its internal characteristics will differ from

A technician at the National Ballistics Intelligence Service uses a microscope and computer to compare barrel markings on two bullets. *(Press Association via AP Images)*

those of its predecessor. Because of these minute alterations in the profile of the cutting tool, no two rifled gun barrels can ever be exactly the same. This uniqueness is transferred—in the form of distinctive scratch marks (striations)—to each bullet that passes through that particular barrel. It is these striations on the bullet itself that form the bedrock of modern ballistics study.

Unfortunately, hundreds of years would pass before anyone realized the identification possibilities that rifling afforded. Two pioneers of ballistics research, an upstate New York physician named Albert Llewellyn Hall and Professor Victor Balthazard, at the Sorbonne in Paris, both of whom worked at the beginning of the 20th century, had hinted at what was achievable. When the First World War broke out in 1914, firearms identification was still a fledgling science, distrusted and without any proper research facilities. This was especially the case in the United States. But all that was about to change.

The Birth of Ballistics 29

At around 5:00 A.M. on March 22, 1915, Charles Stielow, a 36-year-old feeble-minded German immigrant, awoke at the farm where he worked in West Shelby, New York. After breakfast, he opened the door of his tenant house to find, sprawled across the verandah, the body of Margaret Wolcott, the farm housekeeper. She had been shot to death. A trail of footprints in the snow led Stielow across the road to the farmhouse, where he found his employer, Charles Phelps, unconscious on the kitchen floor. Although still alive, Phelps would succumb to his injuries just after midday. Autopsies revealed that both victims had died from .22-caliber gunshot wounds. As the farmhouse had been ransacked, police attributed the murders to a bungled robbery.

At an inquest on March 26, Stielow, whose grasp on events was childlike at best, denied owning any .22-caliber weapons. Despite his denial, he could not shake off local suspicion. At this time in Orleans County, serious crime was sparse and there were no full-time detectives on the local police payroll. Whenever a serious crime did come along, it was customary to hire private investigators on a pay-for-results basis. With a sizable reward on the table—Phelps had been a wealthy man—the promise of a big payday brought out the worst in the contract detective team. Their methods were ruthless. In the days that followed, they targeted Stielow's slow-witted brother-in-law, Nelson Green, who also lived at the tenant house. He admitted hiding two weapons—both .22 caliber and both belonging to Stielow. He claimed to have acted on his brother-in-law's instructions. After a brutal interrogation, which included a threat of lynching, Green confessed that he and Stielow had committed the murders.

Stielow was arrested and thrown into the local jail. Over the next three days, teams of investigators took turns grilling him nonstop. After being questioned around the clock, with only scant pauses for sleep or food, he finally admitted owning the two weapons, but claimed he and Green had hidden them only because they knew that a search was on for all .22-caliber firearms. The interrogators capitalized on Stielow's vulnerability and gullibility. They dangled a promise in front of him: Confess to the murders, they told him, and you can go home to your wife. It worked. Worn down psychologically and physically to the point of utter exhaustion, Stielow allegedly admitted the crimes. Significantly, though, despite the purported verbal confession, he resolutely refused to sign any statement to this effect.

At this point, one of the shadiest characters in the history of American jurisprudence entered the proceedings. Albert Hamilton, a pharmacist from Auburn, had been hoodwinking juries for years, most often on behalf of unscrupulous prosecutors and always at considerable profit to himself. He was a shameless self-promoter who managed to exude an air of immense authority on a wide variety of subjects. Unfortunately for Stielow and Green, one of those pseudo-specialties was firearms. Hamilton had skimmed a few European texts on early attempts at bullet identification and had equipped himself with a microscope and camera—sufficient, he reasoned, to add the qualification "gun expert" to his résumé.

If Hamilton was an expert in anything at all, it was amateur psychology; he knew just how impressionable juries of that era could be when bombarded with overblown jargon and intimidating photographic enlargements. After inspecting Stielow's revolver, he claimed to see "nine defects"[1] in the barrel; those same defects, he asserted, had produced distinctive scratches on the bullets removed from Phelps's body. To bolster this claim, Hamilton took several photographs of the bullets.

When Stielow stood trial for his life, the judge was plainly troubled by the circumstances of the alleged confession and made his skepticism known to the jury. He was also concerned about the fact that no trace of the missing money had been found in Stielow's possession. To counter, of course, the prosecution had the testimony of Hamilton. He did not disappoint. Brandishing his photographs, Hamilton loftily declared that the murder bullets had come from Stielow's gun and no other.

Hamilton's testimony puzzled Stielow's counsel, David A. White. He examined the photographs carefully and professed himself unable to spot the telltale scratches on the bullets. Hamilton, with breathtaking temerity, glibly explained that, by mistake, the photographs showed the side of the bullets opposite the scratches. So authoritative was Hamilton's reply that this astounding response passed without challenge.

And there was worse to come. When White asked Hamilton to point out the alleged defect in the revolver's barrel, Hamilton replied that this was not possible. Because the cartridge fitted the barrel so tightly, he said, the explosive gases had nowhere to go but forward. Consequently, the bullet had acquired so much momentum that its lead content had expanded at the muzzle and filled in the defect, thus rendering it invisible.

Today, such nonsense would be laughed out of court, but in 1915, with firearms identification still very much in its infancy, Hamilton's testimony was powerfully compelling. The jury certainly thought so, and Stielow was condemned to death. (In a brief hearing, Nelson Green later pleaded guilty and received life imprisonment.)

While on death row, Stielow's understated, almost apologetic manner made a deep impression on Sing Sing's deputy warden, Spencer Miller. In conversations with the simple-minded laborer, Miller became convinced of Stielow's innocence and referred the case to the Humanitarian Cult, a New York-based organization dedicated to the abolition of capital punishment. Their efforts on Stielow's behalf were tireless and at times harrowing. Three appointments with the executioner were halted only at the final hour; the closest call happened on July 29, 1916, when Stielow came within 40 minutes of being strapped into the electric chair before a stay of execution was phoned through.

Heartened by this success, lawyers working for the Humanitarian Cult redoubled their efforts. They learned that two drifters named Erwin King and Clarence O'Connell, who were already serving sentences for other robberies, were known to have been in West Shelby on the night of the murder and that King had been heard discussing the crime the next morning—before it was public knowledge. Urged by members of the Humanitarian Cult to clear his conscience, King finally confessed that he and O'Connell had committed the double murder.

This revelation did not sit well with the Stielow prosecution team. Having spent thousands of tax dollars on the case, they were determined to make the conviction stick. They duly whisked King from prison and took him into hiding for several days, after which he dutifully recanted his confession. But it was too late. News of King's confession had already reached Governor Charles Whitman in Albany. Although not entirely convinced by King's story, Whitman felt it created enough doubt to warrant commuting Stielow's death sentence. On December 3, 1916, Stielow was removed from the shadow of the electric chair and later transferred to Auburn Prison, where he joined his brother-in-law. Whitman's job was not yet done. He continued to be troubled by the

COMING TO GRIPS WITH THE EVIDENCE

Ballistics expert don't always need bullets to identify a murder weapon. Between June 1955 and March 1959, nine people in the Washington, D.C./Virginia/Maryland area were gunned to death by a serial killer. The killings seemed random and unconnected, and the only clues were two plastic handgrips from the butt of a large handgun such as a .38 caliber that were found close to the bodies of two victims. FBI ballistics expert George Berley thought the grips had been wrenched from the gun by a sudden force, possibly caused by using the gun as a club. Meanwhile, the search for the killer went on.

Finally, in June 1960, five years after the original murders, investigators received a tip that led them to a 31-year-old itinerant musician, Melvin David Rees. When detectives searched the Hyattsville, Maryland, home of Rees's parents they found a .38 Colt Cobra pistol and a diary that contained graphic accounts of the killings, together with related press cuttings. Rees was arrested a short time later in Arkansas.

When Berley examined the .38 revolver found at the Rees household, he removed the bone handle grips and made an impression of the steel butt. Using a comparison microscope, he compared its surface with the discarded plastic grips found at the scene of Carroll Jackson's murder. The screw holes and various scratches and markings on the Cobra butt and the grip corresponded precisely. When Rees stood trial on January 25, 1961, for two of the murders, Berley told the court that no two gun butts could be so alike.

On February 24 Rees was convicted of double murder and sentenced to life imprisonment. The following September he faced similar charges in a Virginia courtroom that brought him a date with the electric chair. However, Rees's frail grip on reality had now collapsed entirely and his death sentence was eventually commuted to life imprisonment. He died in 1995.

case and, in January 1917, announced the formation of a commission to investigate the circumstances of Stielow's conviction.

The commission was headed by George Bond, a Syracuse lawyer who chose as his assistant Deputy Attorney General Charles E. Waite. By their own admission, both men began the inquiry convinced that the courts had returned just verdicts, but after questioning the principals, especially King and O'Connell, doubts began to creep in.

And there was still Hamilton's evidence to overcome. Bond went looking for acknowledged experts in the field of firearms identification. He deliberately steered clear of professional witnesses who might have a vested interest in the outcome of any analysis and, instead, restricted his search to police officers, working on the assumption that they would be unlikely to be prejudiced in favor of any accused person. The obvious place to begin was the New York Police Department (NYPD). Here, by common consent, the two top ballistics experts were Captain William A. Jones of the Third Branch of the Detective Bureau and Sergeant Harry F. Butts. The two officers met with Bond and Waite on May 14, 1917, at the Third Branch, 219 East 116th Street. Also present was Lieutenant James A. Faurot of the NYPD, America's foremost fingerprint expert and a leading light in the campaign to put crime investigation on a more scientific footing.

Over the next two days, Jones and Butts fired several test rounds from the Stielow pistol. Even to the naked eye, these bullets bore no resemblance to those fired in the Shelby killings. Under a microscope the differences were even more apparent. Both Jones and Butt were adamant: Stielow's revolver was not the murder weapon. Only then did Bond permit Jones to clean the weapon. The amount of rust and grime that Jones removed in four separate cleanings only reinforced the doubts. He and Butt agreed that this gun had not been fired in four or five years, maybe, and certainly not since well before the night of the murder. Although this was only opinion and not verifiable fact, it did cast an enormous shadow over Hamilton's claim, made two years earlier, to have made several test-firings with this gun. If he had lied once, thought Bond, maybe he would lie again? It did not take him long to find the answer. Hamilton's pronouncement that

(Continues on page 36)

MATCHING THE BULLET TO THE GUN

What happened to Stielow and Green was grotesque but not entirely in vain. Their ordeal so outraged Charles Waite that he abandoned the law and devoted the rest of his tragically brief life to improving the science of firearms identification. His approach was distinctly hands-on. He contacted hundreds of gunmakers in the United States and Europe, asking for samples of their weapons in order to catalogue them. Waite was a superb organizer but he had one gaping hole in his credentials—he lacked specialized firearms knowledge. This gap was plugged by Major Calvin H. Goddard, the man who is generally credited with having coined the term *forensic ballistics* (although he later disavowed the term in favor of *firearms identification*).

Goddard probably knew more about firearms than anyone else in the United States. In 1925 he wrote an article for the Army Ordnance titled "Forensic Ballistics," one of the first texts to deal with the systematic identification of guns and bullets. And in April of that same year, he and Waite, along with two others, Philip O. Gravelle and John H. Fisher, founded the Bureau of Forensic Ballistics, the world's first crime lab specifically devoted to the study of firearms identification.

Operating out of a tiny laboratory in lower Manhattan, these four men revolutionized the study of firearms. For the first time, bullets could be matched authoritatively to guns. They were able to do this because Gravelle, a chemist, had invented what is still the greatest advance in the field of firearms identification—the comparison microscope. In this device, two microscopes are connected by an optical bridge, which provides a split-view window, enabling two separate objects such as bullets to be viewed side by side. If the striations match, both bullets were fired from the same gun. If they differ, it is overpowering evidence to the contrary. Gravelle claimed that he came up with the idea of the comparison microscope because he mistrusted his memory. Trying to remember the placement of the minute flaws and imperfections on the bullets was beyond him (or anyone else, for that matter). When it came to examining bullets, he wanted to see the crime scene bullets and the test bullets alongside each other. Only that way could he give a definitive answer to the question of whether the bullet was fired from the gun in question.

At around the same time, Gravelle and Fisher also developed the helixometer—a lens attached to a probe that was inserted into the gun barrel, to examine every part of its interior for flaws that might be reproduced on a bullet fired through it. Although much heralded at its inception, the helixometer failed to live up to its early promise and would always be overshadowed by the comparison microscope.

The first high-profile trial to prominently feature Gravelle's groundbreaking invention was the sensational Hall-Mills murder case in 1926. What should have been an unqualified triumph for the Bureau of Forensic Ballistics was marred by tragedy when Waite, who was scheduled to be a witness for the prosecution, died unexpectedly on November 13, 1926, on the eve of his testimony. Goddard took his place. Although the defendants were acquitted, the validity of the comparison microscope was clearly demonstrated and it was not long before Goddard was once again hitting the headlines, this time as a key witness in the sensational saga of Sacco and Vanzetti, two Italian anarchists convicted of double murder. In 1927 Goddard and his comparison microscope convinced even the defense experts that Sacco's gun had undoubtedly been used in the murderous robbery. Shortly thereafter, over worldwide protest, Sacco and Vanzetti went to the electric chair.

The Bureau of Forensic Ballistics was a short-lived venture. In 1929 Goddard moved to Northwestern University in Evanston, Illinois, where he founded the Scientific Crime Detection Laboratory, America's first multidisciplinary crime lab. Under Goddard's editorship, the lab began publishing the *American*

(continues)

(continued)

Journal of Police Science, which dealt with the latest developments in science-based crime detection. An avid reader of this journal was J. Edgar Hoover, director of the Bureau of Investigation (later the FBI), and he would incorporate many of Goddard's recommendations when he set up the bureau's own Technical Crime Laboratory in Washington, D.C., in 1932. By the time of Goddard's death in 1955, thanks largely to his work, the United States had taken the global lead in firearms identification, a lead that has endured to the present day.

(Continued from page 33)

the barrel's tightness did not permit the backward expulsion of gases was shown to be nonsense when Butts held a piece of paper behind the gun and pulled the trigger. A sheet of flame flew back, igniting the paper instantly. By this time Bond was boiling with rage. It was almost beyond belief that Hamilton had been prepared to lie and see an innocent man go to the electric chair, but what other conclusion could be drawn?

Bond next shipped all the bullets to the Bausch & Lomb Company in Rochester for microscopic analysis by Max Poser, a specialist in optics. Even under the highest magnification, Poser was unable to locate the alleged scratches described by Hamilton, on either the test or murder bullets. All of this, of course, could be taken as merely one opinion against another, but then Poser examined Stielow's revolver and made a discovery that conclusively exposed the worthlessness of Hamilton's testimony. It had to do with the rifling inside the barrel. Poser saw that Stielow's revolver displayed a conventional rifling pattern of grooves and lands, whereas the murder bullets had been fired from a gun that had an abnormal rifling pattern. Put in simple terms, the murder weapon had a manufacturer's flaw, and Stielow's did not. Presented with this incontrovertible evidence, Bond's report unhesitatingly recommended that both prisoners be pardoned. On May 9, 1918,

A criminalist with the Iowa Department of Criminal Investigation looks at one of the more than 3,300 guns the agency uses for reference in its ballistics lab. *(AP Photo/Steve Pope)*

Charles Stielow and Nelson Green walked through the gates of Auburn Prison, free men.

King and O'Connell were never tried for the West Shelby murders. After King had once again retracted his confession, the Orleans County grand jury, mindful of limited local enthusiasm and funds for another lengthy trial, declined to return indictments, and the matter was quietly shelved. Albert Hamilton, too, escaped any censure and continued to testify in trials until his death in 1938.

You Are What You Eat

For a city blighted by so much crime, New York City was surprisingly late joining the forensics party. In the United States it was Chicago that led the way in the use of fingerprinting as an identification tool, and it was Boston that designed the template for the modern medical examiner. It was in this latter field that New York really lagged behind. At the beginning of the 20th century, the scientific study of violent death in America's largest city was left solely in the hands of coroners. These were elected officials with no medical training whatsoever. Generally, they obtained the post as a result of political favor. Cronyism and corruption were rampant. At every step of the way the coroner circumvented the law and fattened up his own pocketbook and those of his friends. The same faces cropped up again and again on inquest juries. Undertakers and familiar doctors appeared with regularity whenever an autopsy was required, irrespective of their competence. Even insurance companies got in on the act, always ready to slip a few bucks to any coroner prepared to sign a death off as suicide, thereby sparing them a hefty payout on any life policies. Bodies were valuable commodities. For instance, it was not unknown for a drowning victim to be hauled out of the Hudson, only for the coroner to issue a death certificate, pocket the $11.50 fee, and then heave the body back into the water, and commence the recovery process all over again.

Something clearly had to be done, and when John Purroy Mitchell was sworn in as New York City mayor in 1914, he set out to clean up the mess. He ordered his Commissioner of Accounts, Leonard M. Wallstein, to conduct a root-and-branch investigation of New York's coroner system, to see if all the rumored corruption and excesses were merely the product of partisan politics or if they were grounded in truth. Wallstein's report, published in 1915, amounted to a blistering attack on the coroner system and recommended the establishment of a single medical examiner's office—fashioned after the Massachusetts model—with responsibility for every unnatural death in the five boroughs. Unsurprisingly, this did not sit well with the various coroners, all of whom lobbied to preserve the status quo and the state of their bank balances. But they were fighting a rising tide of change. They held out for almost three years, but on January 1, 1918, the Office of the Chief Medical Examiner (OCME) was established, with headquarters in Bellevue Hospital, thus beginning that establishment's long association with the OCME, a relationship that endures to the present day. After some last-minute wheeling and dealing, the job of chief medical officer was finally given to Dr. Charles Norris.

Norris was a charismatic figure, eager to raise American forensic science to the standards that he had witnessed in Europe. But he knew that many in America were skeptical; some, especially in New York, wanted him to fail. What he needed was a landmark case—something that would demonstrate to the public the marvels of science-based investigation. Such an opportunity came along in 1922.

For years Abraham Becker and his wife, Jennie, had endured a bitter, loveless union plagued with violent quarrels. Becker's job as a chauffeur gave him plenty of time for philandering and he did not waste a second. He was a prolific womanizer. Finally, in 1920, Becker abandoned his wife and four children to elope with his latest mistress, 24-year-old Anna Elias. Three blissful months in Cleveland came to an abrupt conclusion when Jennie suddenly showed up on the doorstep and punches began to fly from both sides. Anna, bitterly disillusioned because she had known nothing of Becker's duplicity, vowed to end the relationship there and then, despite being pregnant. While Becker returned to the Bronx, and the bosom of his family, Anna set up home in Staten Island and tried to get on with her life. But Becker tracked her down.

(Continues on page 42)

A FLAIR FOR FLAMBOYANCE

There has been no more colorful character in the history of American forensic science than Dr. Charles Norris. With his penetrating, deep-set eyes, imposing bulk of more than 200 pounds, and a glossy Vandyke beard that looked as if it had been sculpted into shape, Norris looked more like a Shakespearean stage actor than America's premier forensic pathologist of the early 20th century. The sight of Norris, with his cape and polished cane, swishing up to a crime scene in a chauffeur-driven limousine was startling at first, but there was real substance behind the showy appearance. He was born into a wealthy Hoboken, New Jersey, family on December 4, 1867, and attended Yale University before going on to study medicine at the Columbia School of Physicians and Surgeons.

He received his medical degree in 1892. He was fascinated by pathology but suffered the same frustration as that endured by fellow students on his side of the Atlantic: All the best teachers were in Europe. For most this was an insurmountable hurdle; for Norris, with his bottomless checkbook, it was an opportunity to soak up the sights and splendor of Europe's greatest cities while receiving top-flight instruction. In 1894 he traveled to Germany. There he enrolled for two semesters in Kiel and one in Göttingen, before traveling to Berlin, where he studied under the legendary pathologist Rudolf Virchow. From 1895 to 1896 he collaborated with two outstanding Viennese teachers, Eduard von Hofmann and Alexander Kolisko. The final leg of Norris's forensic jaunt across Europe took him to Scotland, and the university faculties of Glasgow and Edinburgh, home to the Glaisters and the Littlejohns, two academic dynasties that had revolutionized British forensic science. By the time he returned to his homeland, Norris knew more about legal medicine than anyone in the country, and he was eager for a chance to demonstrate his newfound skills.

In 1904 he was offered the post of professor of pathology at Bellevue Hospital and soon began conducting autopsies in criminal cases. The police trusted Norris. They liked his calm, deliberate manner of speech, and the way he delivered testimony, clearly and succinctly with none of the pomposity that afflicted so many so-called expert witnesses.

At the time of his appointment as chief medical examiner to New York City, Norris outlined his goal, saying that he wanted to establish "a medico-legal institute which would do research work along the lines being done in the larger central European cities. There is no reason why a city of the size and magnificence of New York should not do this work."[1]

Above all, the institute had to be professional. Norris was determined to set new standards of crime lab efficiency. One of his first tasks was the introduction of a telephone switchboard, manned 24 hours a day. Whereas in the past coroners had sauntered up to a crime scene, either at their leisure or after they had sobered up, Norris expected a prompt response at any hour of the day or night.

During the Depression, when staff salaries at the OCME were pared down, Norris frequently dug deep into his own pocket to help those struggling to pay bills. It was the same with office equipment. If a new microscope was needed and the coffers were bare, Norris picked up the tab. His generosity made him a hero to his staff. At the same time, he could be ruthless over any hint of incompetence. He was brilliant and he was tough, and he was exactly what the OCME needed at the time. Following his death on September 11, 1935, 30 patrolmen from the New York Police Department formed a guard of honor at his funeral. The *New York Times*, in an editorial, mourned his passing and lauded Norris as "incorruptible and free from political influence of every kind. When he said a man had committed suicide, that was the truth of the matter. When he said that a man had been murdered, even the doubting police usually came to his way of thinking."[2] It was the best epitaph he could have had.

(Continued from page 39)

All through 1921 he pestered his former lover, always vowing that he would leave his wife, but never delivering on that promise. By the spring of 1922 Becker's entreaties took on new hollowness, especially when it became apparent to everyone who knew him that his marriage had suddenly acquired a new and entirely unexpected warmth. For the first time in years, Becker and Jennie stopped fighting and started loving. So dramatic was the improvement in the Beckers' domestic situation that on the night of April 6, 1922, no one was surprised to see the couple enjoying themselves at a friend's party, with Abe playing the attentive spouse, plying Jennie with canapés, grapes, figs, and almonds. Laughing and giggling, the couple left the party at around midnight in Becker's Dodge sedan, heading north toward the Willis Avenue Bridge.

The next day Becker went to the police and reported his wife missing. He claimed that he had gone to work as usual that morning, but that when he returned home at lunchtime, Jennie had gone. He could offer no reason for her disappearance, but was vitriolic about the way she had left him to fend for four children on his own. The police duly filed a report and promptly forgot all about Jennie.

Becker went back to his apartment on 150th Street. Inquisitive neighbors were parried with the story that Jennie had run off to Philadelphia; another man, apparently, had lured her away. Few believed that Jennie, a devoted mother, would ever abandon her offspring. As for Becker, he raised the temperature of local gossip to dangerous levels by dumping his children in local orphanages, thus clearing the way for Anna Elias and their two-year-old daughter, Marie, to move in with him. If his actions were rash, then his tongue was positively suicidal. On one occasion he boasted to a friend, "Congratulate me. I have got rid of my wife."[3]

Such comments led inevitably to his downfall. In November the chauffeur was arrested as a material witness in the disappearance of his wife. Even in jail his tongue refused to stay still, as he bragged to a visiting acquaintance that he had killed his wife and "buried her so deep they [police] couldn't find the body in 100 years."[4]

During these conversations Becker made frequent mention of a business associate, Reuben Norkin, an auto shop owner. Norkin was quickly arrested and interrogated. At first he denied the allegations, but then admitted helping Becker to bury his wife, although he insisted that his complicity stopped there. But the questioning was relentless and, eventually, Norkin broke down and admitted the whereabouts of the missing woman. On November 25, 1922, he led detectives to the yard of his auto shop. After four days of digging, the search party uncovered a rotting corpse—the eight-month search for Jennie Becker had come to an end. The corpse was taken to the Fordham Hospital morgue, where the OCME maintained its Bronx headquarters. There, Deputy Medical Examiner Dr. Karl S. Kennard performed an autopsy.

In the meantime, Norkin gave his version of how the murder had taken place. On the way home after the party, Becker, an experienced mechanic, had made it appear as though the car had developed engine trouble. The car had spluttered to a halt outside Norkin's auto shop. Some time later, while Norkin watched from a distance, Becker raised the hood to diagnose the problem, then called for Jennie to help him. As she climbed from the car he bludgeoned her with a wrench, raining blow after blow on her head. Certain she was dead, Becker and Norkin hauled her body to a nearby grave they had already prepared. For added insurance Becker doused the body with lime to speed the decomposition process. Then both men drove home.

When it came to interviewing Becker, detectives tried scaring the truth out of him. They marched him into the mortuary and dramatically yanked the sheet off the rotting remains. Becker did not bat an eyelid. He merely sniffed his dismissal. "My wife's a bigger woman than that."

"What about the clothes?" an officer asked.

"These aren't the clothes my wife was wearing."[5] Becker's insistence that his wife had been wearing low-heeled shoes, unlike the fashionable high heels worn by this corpse, prompted retrieval of the original missing person report he had filed. In it, he had described exactly the clothes worn by this dead woman. Still, he refused to budge from his assertion that these were not the remains of his wife.

To settle the argument once and for all, the district attorney's (DA) office turned to Kennard. The crushed skull went without saying, but the congested state of the bronchial tubes made it appear, to Kennard at least, as though Jennie Becker had actually been alive when buried and had died from suffocation. Also, the lime had not totally destroyed the stomach or its contents. Kennard found a small quantity of undigested food. While this could tell him nothing about how long the body had been in the ground, it did provide a clue as to how long after her final meal Jennie Becker had died.

By and large, digestion is a fairly predictable process. According to Dr. Michael Baden, the eminent pathologist, "Very little interferes with the law of the digestive process. It is not precise to the minute (no biological process is), but within a narrow range of time it is very reliable. Within two hours of eating, 95 percent of food has moved out of the stomach and into the small intestine. It is as elemental as rigor mortis. The process stops at death."[6] However, any number of variables can affect this process. Fatty food takes longer to digest, as does a large meal. Mood, too, can play its part, but nothing is more likely to disrupt the digestive process than sudden trauma, either physical or psychological. Fear, fright, injury, or pain, all of these can hamper digestion. For example, after severe head trauma sustained in a road accident, food may stay in the stomach for several days, looking as fresh as when it was swallowed. These instances, however, are exceptions. As a rough rule of thumb, the average meal evacuates the stomach after a couple of hours or so. Here, it looked very much as if Jennie Becker had been killed shortly after eating her final meal.

Kennard bagged the stomach contents and sent them to Dr. Alexander O. Gettler, the OCME's chief toxicologist. Gettler, at this time, was the foremost forensic chemist on the East Coast, and news of his involvement prompted the DA's office to leak details to the press that hinted that Jennie Becker had been poisoned before being bludgeoned to death.

The overheated press coverage—an estimated 50,000 people would eventually line the streets of the Bronx as Jennie's funeral procession passed through—began to wear on Charles Norris's patience. He called for the autopsy report to be brought to his office. As he read it through, he began to bristle. Something was not right here. Norris decided to take a look for himself at the rotting corpse. He concentrated on the fact that the body had lain, unpreserved, in the ground for eight months. Decomposition and postmortem staining made any observations tricky.

You Are What You Eat

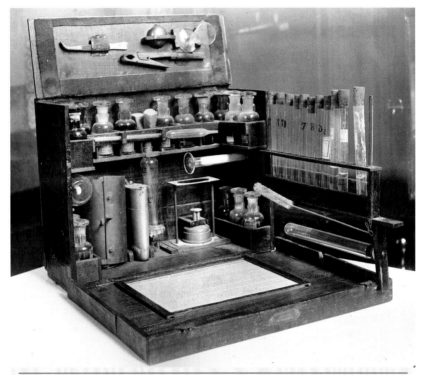

This toxicology kit is an example of the equipment used by crime lab personnel at the time of the Jennie Becker murder investigation. *(Roger Viollet/Getty Images)*

Norris's own examination of the remains left him convinced that Kennard had overstepped the mark. "I don't see any *medical evidence* that suffocation was the cause of death," [italics added] the chief said later. "It may or may not have been."[7] This was a crushing blow to Kennard's prestige. Under the old system, coroners' physicians were used to having their opinions rubber-stamped, not retracted, by superiors. Norris suspected that Kennard had been bullied into his conclusion by a DA's office desperate to wring every inch of emotional mileage out of a killer burying his wife alive. Norris would have none of it. As far as he was concerned, forensic science counted for nothing if the OCME's independence was undermined by meddling politicians. Norris saw what Kennard saw, the same contamination of the bronchial tubes, and he would have defied anyone to say that this provided categorical proof of suffocation. He was equally scathing over press speculation that

Jennie might have been drugged or poisoned, saying, "There has been no evidence of poisoning . . . . No evidence has been found of any wound except the wound on the head."⁸

With Becker still insisting that the remains were not those of his wife, the DA's office could have done without Norris's criticisms. But

THE FUTURE OF THE CRIME LAB

In February 2007 the OCME opened a new high-sensitivity Forensic Biology DNA laboratory at 421 East 26th Street. Costing in excess of $250 million, it is the finest facility of its kind in North America, and the rival of any in the world. Everything is state-of-the-art. A sophisticated laboratory battery-backup power protection system has been installed to ensure that the lab suffers no potentially calamitous power fluctuations of the kind that might damage both crime-related samples and expensive hardware.

A series of rooms is connected by sanitized glass cabinets through which evidence is passed by technicians wearing gowns, gloves, masks, even bootees. Everything is done to ensure that purification levels are the highest possible. Test tubes, instead of being sterilized, which only guards against bacteria, are irradiated to destroy potentially harmful stray chromosomes.

In the past the high cost and time-consuming nature of DNA testing meant it was reserved for only the most serious offenses, typically homicides and sex crimes. This new laboratory's capabilities allow the OCME to assist in burglaries, home invasions, and other crimes. Another useful corollary is that it also significantly increases the number of forensic samples that New York City contributes to the state's DNA database.

When Charles Norris inaugurated the OCME in 1918 he had to suffer cramped, shared facilities with Bellevue Hospital. He could never have dreamt of what the future held for his creation.

any hard feelings that they harbored toward the OCME soon vanished: Gettler had pulled off a forensic miracle. Through meticulous analysis of the stomach contents he was able to find traces of grapes, figs, almonds, and canapés—the very items Becker had lovingly fed his wife at the party. Becker, obviously shaken by these discoveries, blustered that any woman could have eaten such food, until Gettler administered the finishing touch: Examination of meat found in the stomach showed it to be identical to the canapés served by the party hostess—prepared according to an old family recipe.

All his life Becker had lived on his wits. Now his imagination went into overdrive. He finally admitted that the corpse was that of his wife but claimed that Norkin had killed her to "get even with me, because we had a row over an automobile." Asked why he had not reported the murder, Becker replied, "Because he [Norkin] has killed a lot of other people and would kill me, too."[9] Becker then added the bizarre comment that, after killing Jennie, Norkin had considered the account settled, and the two men had become firm friends again.

Predictably, Norkin's version of events was an exact mirror image of the above, even down to his fearing that Becker would kill him, too. Norkin said that Becker had first broached the subject of killing his wife several months before her disappearance. "In April he asked me to lend him a shovel and I let him have one . . . later on he told me that he had buried his wife in a pit near my shop."[10]

While Becker's guilt was beyond dispute, prosecutors opted to seek the maximum penalty for Norkin, as well. In separate trials both men were found guilty of first-degree murder. Becker died in the electric chair on December 13, 1923. The following April it was Norkin's turn to shuffle those last few terrible yards, protesting his innocence with every step.

Thanks to Gettler, this case established and legitimized the OCME in the eyes of the public. Unfortunately, its earliest facilities at Bellevue Hospital were nothing special. Norris and his assistants shared the large, skylit autopsy room on the second floor with the hospital pathologists. Some found the close proximity of the four marble tables claustrophobic; others preferred it, taking the view that so many colleagues so close at hand amounted to having access to a human reference library should an especially tricky situation arise. Although bodies were stored in a room lined with centrally refrigerated storage compartments, a lack of air-conditioning elsewhere in the department meant that sometimes, especially at the height of summer, the OCME could be a foul-smelling place, particularly for distressed relatives who had come in to identify a deceased family member. Despite this, at long last New York City had the forensic detection services that its bulging population so desperately needed.

Atomic Evidence

When most people think of nuclear energy, the first image that springs to mind is a gigantic mushroom cloud soaring heavenward, a terrifying symbol of human destruction on an almost unimaginable scale. But this is only part of the story. Far less well known is the ability of nuclear energy to provide the most powerful and versatile weapon in the forensic analyst's armory. It is called neutron activation analysis (NAA), and it has the capability of identifying trace elements at levels as low as one part in 10 billion, or 1 billionth of a gram. In addition, because of its accuracy and reliability, it is generally recognized as the "referee method" of choice when new procedures are being developed or when other methods yield results that do not agree. And yet, despite this astonishing degree of sensitivity, NAA struggled to become a frontline forensic tool. There was one very good reason for this—for the technology to work, it required access to a research nuclear reactor. And in recent decades such facilities have fallen into disfavor, leading to a significant decline in their number. The effect of this regression was that very few scientists acquired the technical expertise to conduct NAA tests, a situation that clearly posed problems in the area of peer review and, consequently, courtroom admissibility of NAA testimony.

The principle of NAA is far from new. It was first proposed in 1936 when the Swedish-based physical chemist George de Hevesy and his assistant Hilde Levi found that samples containing certain rare earth elements became highly radioactive after exposure to a source of

Using nuclear reactors, crime lab technicians are able to detect microscopic levels of elements through a process known as neutron activation analysis. *(Press Association via AP Images)*

neutrons. From this observation, they quickly recognized the potential of measuring nuclear reactions to identify both the type and the amount of the elements present in the samples. In the 1950s scientists became excited over the possibility that NAA might play a role in forensically analyzing trace evidence from crime scenes. But in order to test this theory, they needed an especially complex case, one well beyond the capabilities of ordinary chemical analysis to solve. Such a case came along in the summer of 1958.

The small Canadian city of Edmundston lies on the New Brunswick panhandle where it juts into neighboring Maine. Separating the two countries is the St. John River, and spanning that river is the Edmundston–Madawaska Bridge, which serves as the gateway between the United States and its neighbor to the north. In the 1950s there was less regard for international borders, and people flowed in both directions with hardly any hindrance. One person who made the trip often was 16-year-old Edmundston native Gaetane Bouchard. She was a normal, high-spirited teenager and she loved to go shopping. So no one thought anything of it when, late one May afternoon in 1958, she suddenly announced that she was going out to pick up a few things. When she had still not returned home by 8:00 P.M., her father, Wilfrid—by now gravely concerned—began telephoning Gaetane's friends. They were hesitant in their replies, but one name did crop up a few times: John Vollman, a 20-year-old part-time saxophonist, who lived just across the border in Madawaska. He had met Gaetane some months earlier at a local dance, and since then the couple had enjoyed a casual relationship.

Bouchard immediately drove to the address in Madawaska that he had been given. There he learned that Vollman was at the newspaper printing plant where he worked the night shift. Eventually the two men came face-to-face. Vollman professed utter ignorance of Gaetane's whereabouts. He admitted dating Gaetane at one time but said he hadn't seen her in several months, not since becoming engaged to another girl.

Bouchard got back in his car and drove home. There was still no sign of Gaetane. When 11:00 P.M. came and went with no word, he finally called the police. They duly noted the details, but were less than enthusiastic; after all, fewer than eight hours had elapsed since Gaetane's disappearance. In all likelihood, she would turn up safe and sound. A good scolding was the recommended punishment in such cases. But Bouchard wasn't about to sit back and do nothing. He climbed back in his car and resumed his own search. His immediate destination was a disused gravel pit just outside of town. Friends of the missing girl had reluctantly hinted that this was a popular spot with couples in cars, and that Gaetane was no stranger to its appeal. Tonight, though, the gravel pit lay deserted. Bouchard nosed his car slowly through the darkness, parked, and began exploring the pitch-black surroundings, flashlight in hand. Within minutes its probing beam picked out a suede shoe, instantly recognizable as Gaetane's. Wilfrid Bouchard hardly dared take another step. But he did. And moments later his flashlight's ray fell on the lifeless body of his daughter.

LIGHTING THE WAY

Spectroscopy is one of the earliest and most powerful of all trace evidence analysis tools. It was developed in 1854 when two German scientists at the University of Heidelberg, Robert Bunsen (inventor of the Bunsen burner) and Gustav Kirchoff, began delving into the mysteries of spectrum analysis.

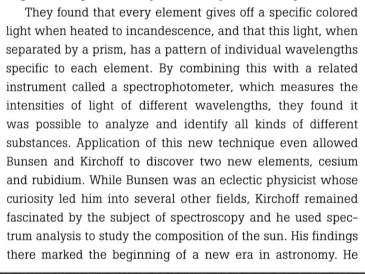

They found that every element gives off a specific colored light when heated to incandescence, and that this light, when separated by a prism, has a pattern of individual wavelengths specific to each element. By combining this with a related instrument called a spectrophotometer, which measures the intensities of light of different wavelengths, they found it was possible to analyze and identify all kinds of different substances. Application of this new technique even allowed Bunsen and Kirchoff to discover two new elements, cesium and rubidium. While Bunsen was an eclectic physicist whose curiosity led him into several other fields, Kirchoff remained fascinated by the subject of spectroscopy and he used spectrum analysis to study the composition of the sun. His findings there marked the beginning of a new era in astronomy. He

She had been stabbed repeatedly in the chest and back. Some distance away a dark pool of blood and a cluster of tire prints signposted where the murderous assault had begun. After a frenzied attack, the killer had dragged Gaetane off to die in the darkness. Although murder was exceptionally rare in this part of the world, the crime scene was processed with a thoroughness that many larger forces might have envied. And it was this meticulousness that uncovered a vital clue. While making plaster casts of the tire prints, an observant police officer noticed two slivers of green paint, one barely larger than the head of a pin, the other somewhat bigger and heart-shaped. They appeared to be from a car, possibly chipped off by flying gravel as the wheels were accelerated fiercely. The evidence was bagged for future analysis.

Atomic Evidence 53

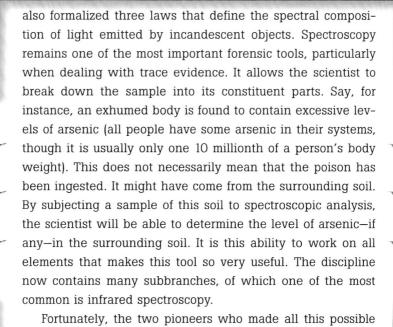

also formalized three laws that define the spectral composition of light emitted by incandescent objects. Spectroscopy remains one of the most important forensic tools, particularly when dealing with trace evidence. It allows the scientist to break down the sample into its constituent parts. Say, for instance, an exhumed body is found to contain excessive levels of arsenic (all people have some arsenic in their systems, though it is usually only one 10 millionth of a person's body weight). This does not necessarily mean that the poison has been ingested. It might have come from the surrounding soil. By subjecting a sample of this soil to spectroscopic analysis, the scientist will be able to determine the level of arsenic—if any—in the surrounding soil. It is this ability to work on all elements that makes this tool so very useful. The discipline now contains many subbranches, of which one of the most common is infrared spectroscopy.

Fortunately, the two pioneers who made all this possible have not been forgotten. Each year in Germany the Bunsen-Kirchoff Award is given for outstanding achievements in the field of analytical spectroscopy.

The next day attempts were made to retrace Gaetane's movements in the hours leading up to her disappearance. At 4:00 P.M. she had bought some chocolate from a local restaurant. Shortly after this she was seen chatting and laughing with the driver of a light green Pontiac, possibly a 1952 model, with Maine plates. An hour later two friends saw her inside a green Pontiac. This was the last time that anyone saw Gaetane alive. Another possible sighting of the vehicle came from a witness who saw a green car—there was no identification of the make or model—parked by the gravel pit between 5:00 and 6:00 P.M.

This timing was significant as the medical examiner who performed the autopsy found partly digested chocolate still in the stomach. From this—and factoring in other variables such as body temperature and rigor mortis—he estimated the time of death at no later than 7:00 P.M.

The only comfort he could offer the bereaved family was that there was no evidence of sexual assault.

For some reason, Bouchard had neglected to mention Vollman's name to the police when he reported Gaetane missing. They learned of him only by talking to friends of the dead girl. And they had plenty to tell. Vollman had a bad reputation. Gradually a picture emerged of an aggressive small-town gigolo, someone used to getting his way, a man who might easily turn violent if any woman rejected him. Then came the zinger: Someone mentioned that, just recently, Vollman had purchased a light green 1952 Pontiac.

Once again Vollman was at work when visitors came to call. Only this time, the callers were detectives. Before interviewing him they had checked the car lot. Vollman's Pontiac was in good condition except for a heart-shaped blemish just beneath the passenger door where the paint was chipped off. When checked, the particle of paint fitted perfectly, a match later verified microscopically.

Vollman treated his inquisitors with a disdain that bordered on rudeness. He did not budge from his insistence that he had not seen Gaetane for several months. As for those eyewitness accounts that placed her in a green Pontiac, Vollman just laughed. If that was the case, he scoffed, then perhaps the Edmundston police chief ought to be interrogated as well, since he drove an identical vehicle. Even the heart-shaped splinter of paint failed to ruffle his composure.

Vollman's cockiness only doubled the detectives' resolve. Armed with a search warrant for his Pontiac, they combed every inch of the vehicle. Inside the glove compartment was a half-eaten bar of lipstick-stained chocolate, just like the type bought by Gaetane a few hours before her death. Although this was strong circumstantial evidence, it was nowhere near enough, in isolation, to convict Vollman. No, they needed something much more compelling. The key piece of evidence came from the Edmundston morgue.

A second autopsy had revealed a most interesting development. Entwined in the dead girl's fingers was a single hair, approximately two and a half inches in length, most probably pulled from the killer's head as she fought for her life. Microscopically there was no way of telling if this hair had come from Vollman's head. Even today, all an analyst can state is that a crime scene hair and a hair from a suspect are microscopically indistinguishable, not that they are identical. Tying this hair to Vollman would require something else, something revolutionary.

Neutron activation analysis is a complicated technique whereby the sample is first placed in a capsule, then inserted in a nuclear reactor and bombarded with neutrons in order to make it radioactive. By measuring the rate at which radioactive atoms disintegrate, it is possible to identify that sample's trace elements. At the time, Canada's premier advocate of NAA was Professor Robert E. Jervis. Since obtaining his doctorate in 1952, Jervis had worked at the Chalk River Nuclear Laboratories, with a goal of finding nonviolent uses for nuclear energy. While there, he became intrigued by the possibilities of using NAA as a forensic tool.

Jervis's faith in NAA was soundly based. Neutron activation analysis works because radioactive elements emit three kinds of radiation: alpha particles (helium atoms), beta particles (electrons), and gamma rays (X-rays). The kind of radiation emitted, and its exact energy, is unique to an element, and can be measured by use of a scintillation counter, a device that is triggered by a flash of light. The method can be used to identify extremely small traces of elements, and their proportions, in various materials.

About 70 percent of the elements have properties suitable for measurement by NAA. Even those elements that are not radioactive—such as sodium, magnesium, zinc, arsenic, cobalt, or potassium—can be made so by being bombarded with subatomic particles called neutrons. The sample is placed in a capsule and inserted into the reactor. When the nucleus of an atom is hit by a neutron, it will often absorb the neutron and become a radioactive isotope of that atom. As a radioactive isotope, it then emits gamma rays, which can be measured to determine the identity, presence, and levels of the original trace element in the sample.

It was at about the same time of the Bouchard murder that Jervis took up a position at the University of Toronto to pursue his NAA studies. He was asked to examine the crime evidence and give his opinion. Jervis took hairs from Vollman and Gaetane and compared them to the single hair extracted from between Gaetane's fingers. In the proportion of sulfur to phosphorus radiation, Gaetane's hair registered 2.02; while the sample from Vollman and the single hair were 1.07 and 1.02, respectively. Evidently, the single hair had not come from the murdered girl, and, just as obviously, it was very similar to that of Vollman. This was enough for officers to arrest Vollman and charge him with murder, but there was no guarantee that Jervis's evidence would ever make it into court. When Vollman stood trial for murder in Edmundston on November 4, 1958, his counsel, J. A. Pichette, argued that such a radically new technique was untested and therefore should not be admitted into testimony. After lengthy deliberation, Judge Arthur L. Anglin disallowed the objection. Vollman, who had originally pleaded not guilty, then had to sit and listen as a string of scientists entered the witness box and validated Jervis's methodology. As they did so, the mood of the court, which had originally been skeptical about what some sections of the press sensationally dubbed "atomic evidence," palpably changed. Vollman could sense this shift. Suddenly, through his counsel, he announced that he wished to change his plea to one of guilty to manslaughter. He now admitted killing Gaetane, but said it had been unintentional. They had driven to the gravel pit and parked. At first, he said, Gaetane had encouraged his advances, but then she had changed her mind. An argument broke out, followed by a struggle. Vollman claimed this was his last memory. Everything after that was a blur, lost in the murk of a mental blackout. Notably lacking from his account was any explanation of how a knife happened to be in his hand at the time of this spontaneous argument erupting.

Vollman's last reckless gamble failed. The jury did not believe him. He was found guilty as originally charged and sentenced to the gallows. Because Vollman was an American citizen—and a resident of Maine, which had abolished capital punishment in 1887—it was always unlikely that he would be executed, and so it proved. On February 14, 1959, four days before his scheduled date with the hangman, his sentence was commuted to life imprisonment, to be served at the maximum security Dorchester Penitentiary in New Brunswick, Canada.

For the first time in history, NAA had been used in a murder trial to gain a conviction. In the aftermath of the Vollman trial, almost anything seemed possible. A string of high-profile cases involving so-called atomic evidence catapulted the new technology temporarily into the forensic spotlight. For example, NAA was used to analyze samples of hair from Napoleon, to investigate the possibilities of arsenic poisoning,

"SEE YOU" IN COURT

Almost any branch of technology can provide forensic evidence, and this certainly applies to the cell phone. According to CTIA—the International Association for the Wireless Telecommunications Industry—in 2009 there were an estimated 276,610,580 cell phones in use in the United States. Before long the number of phones will exceed the number of citizens.

The radio footprints left by cell phones are becoming an increasingly common feature of trials as unwitting criminals use cell phones during illegal activities. Not only can telecommunications experts establish where a handset was used, they can also uncover deleted text messages. Generally, cases involving phone threats, marital discord, or calls made by bragging offenders are among the most likely to rely on phone analysis, but there have been murder trials in which text messages have proved crucial.

A new system known as Celldar adds yet another crime-fighting dimension to cell phone technology. It works by using receivers attached to cell phone masts. These "see" the shapes made when radio waves emitted by cell phone masts meet an obstruction. Signals bounced back by immobile objects, such as walls or trees, are filtered out by the receiver, thus allowing anything mobile, such as cars or people, to be tracked. Users are able to focus in on areas even hundreds of miles away, and bring up a display showing any moving vehicles and people. An individual with a portable unit little bigger than a laptop computer could even use it as a "personal radar" system covering the area around the user. Researchers are working to give the new equipment X-ray vision (the ability to "see" through walls and look into people's homes).

Obviously such intrusive capabilities will cause concern and outrage in certain quarters, but the likelihood is that cell phone technology will play an increasingly important part in the fight against crime.

and to identify the trace elements in bullet lead from the assassination of President John F. Kennedy. But as nuclear energy fell victim to public concerns about safety and the number of reactors in the United States plummeted, there was a very real danger that NAA would fall out of use.

What was needed was a less expensive—and less controversial—means of producing neutrons. A variety of alternatives eventually surfaced, including fusors (devices that use "high temperature" ions to create nuclear fusion), isotope sources such as beryllium, and gas discharge tubes. None is as efficient as a nuclear reactor at generating neutrons, but all have their uses.

Forensic applications for NAA include identifying all kinds of samples typically found at crime scenes—gunshot residues, bullet lead, glass, paint, soil, and hair, for example. Research suggests it may also be possible to identify paper by batches. One other interesting development is in the always-controversial field of determining whether an individual has recently fired a gun. For years, law enforcement agencies relied on the dermal nitrate test, in which melted paraffin is used to pick up traces of nitrates and nitrites on the hand, after which a chemical test is made for such compounds. Following a string of disastrous false positives generated by the dermal nitrate test in the 1990s, it has now been thoroughly discredited. Use of NAA to detect the presence on the hand of antimony and barium from cartridge primers is much more definitive.

One other area where the use of NAA might be of enormous benefit concerns the identification of elements in bullet lead. For years it was generally believed—and widely testified to in court—that the chemical makeup of the lead used to make a particular batch of bullets was peculiar to that batch alone. Now, however, there is evidence to suggest that this may not be the case. Researchers working at a variety of smelters across the United States have found small but measurable differences in the composition of lead samples taken at the beginning and end of the same batch, probably due to oxidation of the trace elements. This makes it impossible to say whether any two bullets were made on the same day or come from the same box. Further analysis only added to the confusion. Using the FBI's chemical profile standards, it was impossible to distinguish between batches poured months apart.

Such anomalies clearly require the most rigorous investigation and highlight the need for the most sophisticated chemical analysis possible. To date this sophisticated analysis is NAA. Apart from its high-level identification capabilities, it has many other advantages. It can analyze the smallest of samples, does not destroy the sample, and can run large numbers of samples at one time. Nowadays even the previously prohibitive cost has been reduced to an economic level. This has prompted a surge in NAA usage. Worldwide, it is estimated that approximately 100,000 samples (mostly geological minerals) undergo analysis each year, though applications in the fields of archaeology and biochemistry are on the increase.

A Disputed Document

In the world of art and antiquities, provenance is everything. With it, an item might be worth millions of dollars; without it, the value could plummet to zero. Provenance is, simply, proof—preferably in documentary form—that an item is what it claims to be, and that its ownership is legitimate. When the world looked on in November 1922 as the English archaeologist Howard Carter unsealed the tomb of Tutankhamun in Egypt's Valley of the Kings, there could be no doubt that the phenomenal riches uncovered had, indeed, come from the boy king's funeral chamber. Because Carter meticulously catalogued every find, both on paper and in photographs, the provenance of each astounding item was unshakeable. Imagine now, if a "previously undiscovered" gold amulet, purporting to be from Tutankhamun's tomb, suddenly appeared on the world's antiquity markets. Such an object would be laughed out the door. This is because it would have no provenance. The amulet might well be 3,500 years old, but it could just as likely have been manufactured in a Cairo back street a few months previously. Unfortunately, very few priceless artifacts from the past are as well documented as Tutankhamun's treasures, and this is where problems arise.

Ever since medieval times, the antiquities/art market has provided an irresistible lure to forgers. Indeed, fakery is probably as old as art itself. When one considers the staggering sums of money involved, this is scarcely surprising. For instance, in 2004 Picasso's *Garçon à la Pipe* sold for $104.1 million, the highest price ever paid for a painting

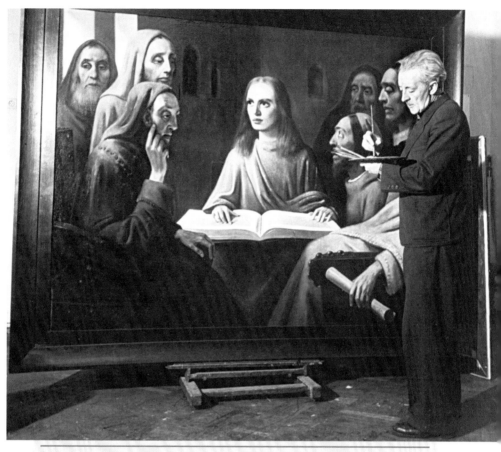

Demonstrating his craft, Hans van Meegeren works on a reproduction of Jean Auguste Dominique Ingres's *Jesus Among the Doctors* before his trial in 1947. *(Time & Life Pictures/Getty Images)*

at auction. Numbers such as this can exert a powerful influence on the criminal mind. As a result, the art and antiquities world is riddled with fakes. Some are so convincing that they have even fooled the experts. When Hans van Meegeren was arrested after World War II and charged with selling a Vermeer painting to the Nazis, he protested his innocence on the grounds that he had painted the picture himself. To prove his point, van Meegeren duly painted another "Vermeer" before an astounded panel of watchers. Other "Vermeers" painted by van Meegeren were discovered in Dutch galleries, all of which had thought they were buying a genuine piece.

IS IT A FORGERY?

Determining whether a document has been altered is usually possible through a process called electrostatic detection apparatus (ESDA). A document is placed on top of an electronically charged metal mesh and a thin plastic film is pulled tightly across it. As the document and the film are sucked tight onto the mesh, a mixture of photocopier toner and fine glass beads is applied, which clings to the electrostatically charged areas. When the original document is removed, all of the indentations on the film may be read. By matching all suspect original pages against each image, investigators can determine whether changes have been made.

Questions concerning the sequence of writing—for example, is the ink signature over or under the typewritten portion of a document?—can be vital in proving the authenticity of a document. In this and related problems, such as deciphering and restoration of eradicated or erased writing, most examinations are nowadays carried out primarily using computer-based technology.

The latest weapon in the FBI's questioned documents arsenal is the Video Spectral Comparator 5000 (VSC5000), an extraordinarily sophisticated imaging device that allows an examiner to analyze inks, visualize hidden security features, and reveal alterations on a document. Using the VSC5000's infrared radiant energy source and filters, the examiner is able to see through inks to reveal objects that are obscured to the naked eye. But it has a variety of other uses.

The specimen is first subjected to a variety of light sources, while cameras and filters record the effect on a VDU. Software enables the examiner to manipulate the image for easier viewing. It allows images to be overlaid or compared side by side, which is useful in performing torn-edge comparisons on sheets of paper, and in determining whether a page has been added or replaced in a multipage contract. It can even aid the examiner to sort shredded pieces of paper, allowing for the reconstruction of a readable document.

Van Meegeren's success demonstrates how dangerous (and expensive) a lack of provenance can be. When it comes to detecting art forgeries, the modern crime lab is far more adept than its predecessors. Advances in analytical techniques, coupled with a greater awareness of the forger's abilities, make it increasingly difficult for the modern faker. It is the job of science to deal in facts; how those facts are interpreted is another matter altogether. Occasionally, however, something surfaces that seems to defy analysis, leading even the experts to scratch their heads and wonder—did a crime take place or not?

In 1965 a document came to light that threatened to turn history upside down. Drawn on a single sheet of parchment, measuring 11 by 16 inches and folded down the middle, it purported to be a medieval map of the Old World, as it was thought to be in the 15th century. In the western Atlantic it depicted a large island called *Vinilanda Insula* (island of Vinland, land of vines) the Vinland of the Norse sagas, and what is, unmistakably, the northeastern coast of North America.

The map was bound to a slender tract of Medieval Latin text entitled the *Tartar Relation* and another volume called *Speculum Historiale*,

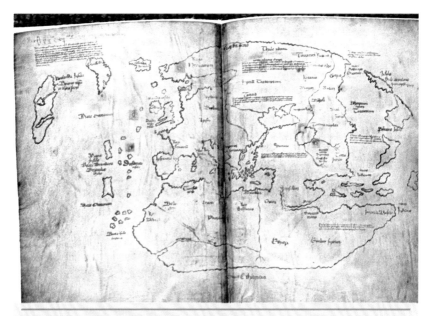

The Vinland Map's authenticity is still a source of debate. The most likely truth is that the parchment is genuine and the ink is fraudulent. *(AP Photo)*

both of which were reliably dated to c.1440. These told how a Viking mariner named Leif Eriksson had traveled to this new land around the year 1000. It is commonly accepted that the Vikings had indeed reached the North Atlantic coastline of America at around this time. But what the Vinland Map, with its Latin inscriptions in brown ink, provided was tangible evidence that Western Europeans—not just the Vikings—were aware of the New World a full 50 years before Christopher Columbus made his epic voyage of discovery in 1492.

It was the kind of historical find to make any antiquarian swoon, and by the mid-1990s even the most conservative estimate valued the map at $20 million. Such valuations—and the eye-watering insurance premiums that accompany them—are strong tests of any buyer's nerve. But one big question has dogged the Vinland Map ever since its discovery—is it a rare and unique medieval masterpiece, or is it a well-made 20th century fake?

It originally surfaced in 1957, when it was offered to the British Museum by an unspecified source. Uncomfortable with the map's murky past and certain irregularities with the binding, the museum had rejected the item. Later that same year a Connecticut antiquarian bookseller, Laurence C. Witten, paid $3,500 for the map. He bought it from a Spanish dealer, Enzo Ferrajoli de Ry, who claimed that the map had been in an unnamed family's possession for two generations. Despite this vague background, Witten was convinced he had stumbled across something of huge importance, and his enthusiasm was infectious. In 1959 the wealthy philanthropist Paul Mellon bought the map from Witten for an undisclosed sum (rumors put the figure at between $300,000 and $1 million). Mellon reportedly subjected the map to a string of authentication procedures that took several years and rang no alarms. In 1965 he donated it to the Beinecke Rare Book and Manuscript Library at Yale University. Eager to wring out every last bit of publicity from their triumph, Yale waited until three days before Columbus Day, 1965, before revealing its latest acquisition. Excited curators characterized the Vinland Map as "the most exciting cartographic discovery of the century."[1] Contemporaneous archaeological discoveries of a Norse settlement in Newfoundland only heightened the map's relevance.

It did not take long for the doubters to surface. In 1966 a Russian, Vladimir Nevsky, a lecturer in geography at Leningrad University, denounced the map as "a fake and rather a too clever one."[2] Nevsky and his colleagues thought the map showed obvious signs of having drawn on stylistic quirks used by Arabic cartographers in the 10th and 12th centuries.

Others questioned the depiction of Greenland as an island. On another, undisputed, map from 1427, Greenland is shown as the end of a peninsula stretching toward the Arctic north. Pack-ice conditions made sailing along Greenland's northern coast treacherous, if not impossible, and the first circumnavigation of Greenland was not completed until the turn of the 20th century. Also, the outline of the island's northern coast bears a disturbing similarity to that shown on modern maps.

Yet another suspicious factor involved the map's fold from top to bottom. Geographic place names have been placed either side of the fold, instead of crossing it. Skeptics wondered at the cartographer's prescience in avoiding a fold that would not have been there when he made the map.

All of these legitimate concerns raised doubts about the map's authenticity. No institute of major learning likes to have its credibility undermined, and Yale was especially sensitive to the criticism. Since taking possession of the map, the university had sold 10,000 copies, at $15 each, of a lavishly illustrated book dealing with the map and its related items. This was embarrassing enough, but hovering in the background was the very real and distinctly unpleasant possibility that a major crime had taken place.

In 1972 Yale, to its credit, decided to see if science and the laboratory could put this argument to rest. It commissioned a Chicago chemist named Walter McCrone to microscopically analyze the map. McCrone found that whoever drew the map first used a yellow ink, then carefully superimposed a black ink line over this. The black outer layer was actually flaking off, but the yellow ink still adhered firmly to the surface, and it was this yellow ink that raised doubts. In it, McCrone detected the presence of anatase, a form of titanium dioxide. Anatase does occur in nature, but only in small amounts and then in jagged, irregular crystals; not the round, uniform crystals that McCrone found here. So far as McCrone was concerned, this form of anatase could have originated from only one source—modern ink. In 1923 ink manufacturers began adding just such uniform crystals of anatase to their products; this was evidence enough for McCrone to declare the Vinland Map a 20th-century fake.

HOW OLD IS THIS OBJECT?

Whenever forensic archaeologists or crime labs are asked to determine the age of an object, they generally employ a technique called radio carbon dating. This is the most widely used method of age estimation and works by measuring the amount of carbon-14 remaining in an object. The principle was pioneered by Willard F. Libby, working with items of known age, at the University of Chicago. Libby's groundbreaking research earned him the 1960 Nobel Prize for Chemistry.

Certain chemical elements have more than one type of atom and, where this occurs, different atoms of the same element are called isotopes. Carbon, which forms more compounds than all the other elements combined, has three main isotopes: carbon-12, carbon-13, and carbon-14. Of these isotopes, carbon-12 is the most abundant, making up 98.89 percent of an atom; next comes carbon-13 at 1.11 percent; while down at the bottom of the scale is carbon-14, which makes up just one part per million. What makes carbon-14 so useful is the fact that, alone among these three isotopes, it is radioactive, and it is the gradual decay of this radioactivity that scientists use to measure age.

Radioactive atoms decay into stable atoms by a simple mathematical process. Half of the available atoms will change in a given period of time, known as the half-life. For instance, if 1,000 atoms in the year 2000 had a half-life of 10 years, then in 2010 there would be 500 left. In 2020 there would be 250 left, and in 2030 there would be 125 left, and so on.

Therefore, by counting the number of carbon-14 atoms in any object that contains carbon, it is possible to calculate either how old the object is or how long ago it died. In order to arrive at this number the scientist needs to know two things:

Few could claim to be shocked. As previously noted, antiquarian objects are only as valuable as their provenance, and the Vinland Map's antecedents were so sketchy as to be virtually nonexistent. Despite this

the half-life of carbon-14 and how many carbon-14 atoms the object had before it died. The first part is straightforward: the half-life of carbon-14 is 5,730 years (+/- 40 years). However, knowing how many carbon-14 atoms a test object had before it died is something that can only be estimated; however, the assumption is made that the level of carbon-14 in any living organism is constant, so when a particular fossil was alive, it contained the same amount of carbon-14 as the same living organism today.

Radio carbon dating is highly effective, with dates derived from carbon samples being tracked back as far as 50,000 years. Beyond that, it is necessary to employ potassium or uranium isotopes, which have much longer half-lives. These are used to date ancient geological events that have to be measured in millions or even billions of years.

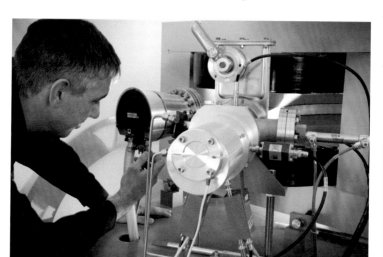

A scientist uses an accelerator mass spectrometer to conduct radio carbon dating of a specimen. (*James King-Holmes/Photo Researchers*)

lack of background information, several experts had staked their reputations on its authenticity.

Yale waited until it had McCrone's findings verified independently before going public with the humiliating revelation. Eight and a half years after taking possession of the disputed document, the university gloomily announced "that the famous Vinland Map may be a forgery."[3] Significantly, Yale decided not to pursue criminal charges against anyone, nor did it offer any refunds to those persons who had purchased the copies of the book. *Caveat emptor* (buyer beware) seemed to be the order of the day.

With so many reputations on the line, academics and scientists lined up to take potshots at McCrone's findings. As the attacks began to mount, even Yale began to soften its stance on the map's authenticity, shifting to a position of neutrality. McCrone, a scientist of profoundly skeptical views when it came to miraculous finds from history, was a natural lightning rod for criticism. In 1979 he hit the headlines once more when he debunked the Turin Shroud as a fake. Some began to wonder if this was a scientist driven more by a desire for publicity than a quest for the truth. There was only one way to find out—it was time to go back to the lab.

In 1995 a team led by Thomas Cahill, a professor of atmospheric science and physics at the University of California, put McCrone's methodology and results under the microscope. Cahill, who had been sniping at McCrone for almost a decade, concluded that most of the crystals McCrone found were not anatase at all, and that a third of the ink contained no titanium.

The cerebral fisticuffs continued to fly. In July 2002 supporters of the map received another boost when a Smithsonian Institute-backed radiocarbon dating of the parchment came up with a date of approximately 1434, well within the time frame of when the map was supposed to have been drawn. After all, said recently retired Smithsonian research chemist Jacqueline Olin, "It's not a trivial thing for a forger to get a parchment [from that time period]."[4] While no one doubted the age of the parchment, radiocarbon testing did nothing to address the all-important age of the ink.

A Disputed Document   69

By coincidence, on the very same day, researchers at University College in London also revealed the results of *their* investigation into the Vinland Map, and these provided unequivocal support for McCrone. Using a Raman microscope, which employs a laser beam that scatters molecules as radiation with different colors, Professor Robin Clark and his colleague Katherine Brown found that anatase was detected solely in the yellowish ink lines, and not elsewhere on the parchment. Since yellow lines are sometimes left behind when medieval ink made of the pigment iron gallotannate degrades, Clark said a forger would know about the yellow residue and try to reproduce it. But the black ink on top of the yellow ink was found to be carbon based, not iron gallotannate, so no yellow residue should be present. Clark stressed the importance of the laboratory when dealing with alleged frauds. "The results demonstrate the great importance of modern analytical techniques in the study of items in our cultural heritage."[5]

The likeliest outcome of all this inter-laboratory sparring is that both sides are correct: the parchment is from the 15th century, and the ink is probably modern. Whoever faked the Vinland Map was clearly an expert in his or her field, and in August 2002, author Kirsten Seaver pointed the finger at a most unlikely culprit.

Father Joseph Fischer, who died in Germany in 1944 at age 86, was an authority on medieval cartography. As a young man it was Fischer who had discovered the first map, dated 1507, to name America—later bought by the Library of Congress for $10 million—and it was Fischer's obsession with achieving firsts, according to Seaver, that prompted the fake. She believes that Fischer succumbed to serious depression after his credentials came under attack in 1934. At about the same time he acquired a volume of 15th-century manuscripts, into which was bound a loose leaf of a piece of parchment dating from the 1440s, and he decided to fool the world.

Seaver claims to have analyzed the record of every other scholar in the field between 1920 and 1957, and insists that the circumstantial evidence points overwhelmingly to Fischer. His handwriting corresponds to that on the map; the incorrect Latinate spellings are what one might expect of a scholar who was proficient in Latin but ignorant of Old Norse; the tract bears the remains of a faded stamp similar to that used by the Jesuit college of Stella Matutina at Feldkirch, Austria, to which Fischer was linked; even the fact that the map showed the world as round hurt Fischer. Seaver says that Fischer was alone in realizing that medieval mapmakers utilized the Ptolemaic knowledge that the world was not flat.

At this point, said Seaver, fraud collided with history. In 1938 the Nazis overran Austria and began to plunder that country's treasures. Fischer was ousted from Stella Matutina and all his works were confiscated. After the war, the Vinland Map reportedly found its way to alleged Nazi sympathizer Enzo Ferrajoli de Ry.

It is quite a story, and entirely unprovable. Seaver has cleverly woven a string of incidents into the kind of tale that would not look out of place in a Dan Brown novel. Whether her allegations stand the test of time remains to be seen, but what her theory does demonstrate is the enduring fascination that the Vinland Map exerts over the academic and scientific mind. Yale has recognized the difficulties. "The truth is there has been a good deal of controversy about the Vinland Map. There is a lot of research in progress," said Yale's former head librarian, Alice Prochaska. "I think probably research will reveal one day what the truth is, but it is certainly very much under discussion and debate."[6]

Given the troubled history of the Vinland Map, such confidence is hard to justify. In the scientific community, an opinion once given is rarely changed and often defended to the death, no matter what evidence might subsequently emerge. As courtrooms around the globe have discovered, getting a lab or a scientist to change their mind is a task of almost Herculean proportions—perhaps even more difficult than pulling off the greatest antiquarian forgery of modern times.

Bags of Evidence

There are few things more frustrating for a detective than being convinced that someone is a killer, yet not being able to prove it. This is one of the main reasons why the modern crime lab came about. Its job is to uncover the evidence that ordinary methods of detection overlook. The hope is that the lab can confirm the investigators' suspicions and track down that vital clue that will nail a heartless killer in a court of law. But evidence cannot (or should not) be manufactured out of thin air; there has to be something that the scientist can analyze. And the truth is that not all crime scenes yield useable clues. Sometimes the circumstances can be so bizarre, so out of the ordinary, as to defy the best that modern science can throw at them. Should this be the case, then the hope is that the murderer will ultimately slip up and convict him- or herself. What nobody wants is another killing. But occasionally, as cruel as it may seem, repetition can be the investigator's best friend.

At about 10.30 P.M. on Saturday, April 29, 1989, a 30-year-old mother named Paula Sims was taking out the garbage from her Illinois home when she was jumped on the back porch by a gun-carrying assailant in dark clothes and a ski mask. At gunpoint he forced her back into the house. No sooner had she stepped into the kitchen than the gunman felled her with a heavy karate chop to the back of her neck. According to Paula, everything then went black. She was still unconscious when her husband Robert arrived home from work at a few minutes past 11:00

JUNK SCIENCE

It is the duty of the crime lab to provide solid, reliable evidence. Sadly, this is not always the case. In 1989 a young Missouri mother, Patricia Stallings, was charged with murdering her four-month-old son, Ryan. Tests showing extraordinarily high levels of ethylene glycol in his blood led investigators to charge Patricia with poisoning her son with antifreeze. At the time of her arrest, Patricia was once again pregnant, and five months later she gave birth to her second son, David. He was placed in foster care. Soon, it became clear that something was amiss with David, and in April 1990 he was diagnosed with an extremely rare genetic disorder called MMA. Methylmalonic acidemia, which affects one out of every 48,000 children, inhibits the body's ability to process food, especially protein. It also produces toxins in the blood. When David was diagnosed with the disease, experts said there was a one-in-four chance that Ryan too had suffered from the rare disorder.

The medical evidence, though, seemed unassailable. First, there was the finding—by two labs—of 911 micrograms of ethylene glycol in Ryan's body. Even if Ryan suffered from MMA, this could not account for the high levels of ethylene glycol. Nor could it account for the autopsy's finding of calcium oxalate crystals in the child's brain and other organs—traces consistent with ethylene glycol poisoning.

In light of these results, Patricia's attorney was unable to find any expert to testify that Ryan might have died of MMA and that he was not poisoned at all, much less by his mother. In February 1991 Patricia Stallings was imprisoned for life without parole.

But help was at hand. Dr. James Shoemaker, a young scientist who had just set up a genetic testing laboratory at St. Louis University—ironically one of the institutions that had

P.M. He tried hard to rouse his wife, but failed. Fearful that the attacker might still be in the house, he decided to search the rest of the house. He made a terrible discovery: Their six-week-old baby daughter, Heather, detected the ethylene glycol—ran extra tests on Ryan's blood that strongly suggested that he had suffered from MMA.

Also, when Shoemaker analyzed Ryan's blood for retention times—the amount of time it takes a compound to travel through a gas chromatography column—he found not ethylene glycol, but proprionic acid, which has a similar retention time to ethylene glycol. Significantly, proprionic acid is often produced by MMA sufferers. Superimposition of a graph over the original database printout showed clear disparities in peaks.

He sent his findings to Dr. Piero Rinaldo, a world-renowned genetics expert from Yale University. Rinaldo was flabbergasted by what he found. He described the previous test results as "unbelievable . . . I couldn't believe that somebody would let this go through a criminal trial unchallenged."[1] In independent tests Rinaldo found no evidence of ethylene glycol in Ryan's blood. And those mysterious crystals found in Ryan's autopsy, he said, were most likely caused by the ethanol drip used in the hospital to treat Ryan's suspected poisoning.

A reexamination of the original data revealed yet another extraordinary blunder—for Ryan to have 911 micrograms of ethylene glycol in his system, he would have had to ingest more than 80 gallons of antifreeze, a clearly absurd amount. There had been no murder, just bad science. Fortunately, better science was at hand and Patricia Stallings was freed and all charges against her were dropped.

It is a sobering thought that had Patricia Stallings not been pregnant, and had David not exhibited the symptoms of MMA, she would probably still be behind bars. This highlights the need for constant vigilance. As courts come to rely ever more on crime labs, the quality of their work must be scrupulous and beyond reproach. Nothing less is acceptable.

was missing. When Robert returned to the kitchen, he tried once more to wake Paula and this time succeeded. She told him what had happened. He gave her the grim news that Heather was missing. Together the couple ran to the bedroom of their 15-month-old son, Randy. He was sleeping soundly, unharmed.

Detectives summoned to the house in Alton, Illinois, were immediately suspicious. For someone who had been unconscious for close to 45 minutes, Paula seemed remarkably unscathed and very lucid. Also, there was no sign of an intruder; no unusual fingerprints in the house; nothing had been taken; and Heather's bassinet and the area around it lay undisturbed. Weirdest of all: Why would anyone want to kidnap the child of a family so patently unable to pay a ransom? There was, of course, the possibility that Heather had been taken by an unhinged person or persons desperate to have a child of their own, but investigators did not set much store by this theory. Instead, they treated the case as a straightforward kidnapping.

The following Wednesday evening, some two miles away from the Sims's residence, a man named Charles Saunders was fishing the Mississippi River, on the Missouri side. He did not stay long, just half an hour or so, then decided to head home. Before leaving, he topped up the oil in his car, then walked over to a rest area to dispose of the container in a trash can. At the bottom of the trash barrel lay a large black plastic bag. It looked unusually bulky. Saunders was curious. One peek inside told him that he had found Heather Sims.

When Detective Mick Dooley, an experienced officer, studied the well-preserved remains, his first instinct was that Heather had been killed recently. But then he noticed that lividity—the tendency of blood to settle in the lowest extremities of the body—which usually becomes fixed some eight to 12 hours after death, had stained the front of her face a dark red, and yet Heather was found lying on her back. Also, there was a bright red mark on the baby's cheek, such as might be caused by freezer burn. His immediate conclusion was that whoever had killed Heather had first stored the body in a freezer before dumping it in the garbage bin shortly before it was discovered.

Medical examiner Dr. Mary Case confirmed Dooley's suspicions about the freezing. During the autopsy, she found that Heather's internal organs showed early signs of decomposition "inconsistent with the exterior of the body."[2] But it was the examination of the mouth that revealed the cause of death. There were lacerations on the inside of the lips, leaving Case in no doubt that Heather had been suffocated.

Bags of Evidence 75

It was a terrible tragedy, one that no family should have to endure. But what astonished, even horrified, the detectives was that Paula Sims, the grieving mother, had gone through a virtually identical experience three years before.

Just about the only difference was the location. Paula and Robert Sims had been living in nearby Jersey County on June 17, 1986, when, according to Paula, a masked intruder had broken into her house at gunpoint and snatched her 12-day-old baby, Loralei. Once again it was after 10:00 P.M. and Robert had been at work. On this occasion Paula had been unharmed and had given chase, only for the kidnapper to elude her in the darkness. The only sign of a forced entry was a four-inch tear in the screen door, near the handle. But investigators noticed that something was not quite right. The screen door did not fit its frame well, and the act of opening it caused it to scrape on the floor, making a loud screeching sound that resonated through the house. And yet Paula, who said she was sitting in a chair watching the news on TV, claimed to have heard nothing before the intruder suddenly appeared at the foot of the stairs. It was certainly curious.

One week later the search for Loralei came to a dreadful end. Her skeletal remains were found in a thicket, just 150 feet behind the Sims's residence. Why a kidnapper would run off with the baby, then return to the crime scene to dump his victim was just one question that vexed investigators. Another was the fact that, due to decomposition and the ravages of wild animals, it was impossible to give a definite cause of death. Despite deep misgivings, detectives realized that they did not have enough evidence to make a case against Paula, so no charges were leveled.

And now, three years later, it had happened all over again. Was it really possible that some phantom kidnapper was targeting the Sims family and that lightning had struck twice? Investigators did not think so for one minute, nor did Dr. Case. Besides being a medical examiner, she was also a specialist in neuropathology, and understood the effects of head trauma on the nervous system. A blow hard enough to render Paula unconscious for almost an hour would cause the brain to collide with the skull. Such a concussion generally wipes out all memory of the incident, yet Paula was able to clearly recount every detail of the attack. This did not make any medical sense. The only logical conclusion was that, for some reason, Paula Sims set out to willfully kill her daughters. But why?

As detectives probed the relationship between Paula and her husband, they uncovered some deeply disturbing allegations. Friends of the family and nurses at the hospital where Paula had given birth on three occasions were united in their opinion that Robert Sims appeared to be utterly disinterested in his baby daughters. By contrast, he doted on little Randy, who always had the best of everything. An even more sinister edge crept in, with suggestions that twice Paula had come back from the maternity hospital with a little girl, and on each occasion Robert had banished her from the marital bedroom, forcing her to sleep with the infant. Only when the baby was dead, reportedly, was Paula welcomed back into his bed.

Such behavior—if true—might be grotesque, but it was not proof of murder. What the investigators needed was hard evidence. Loralei's death had yielded too few clues to press charges; this meant that any hopes of bringing Paula Sims to justice lay in proving that she had killed Heather. Detectives worked up a murder scenario. On the Saturday, while her husband was at work, Paula had suffocated Heather and had then driven across town to her parents' house, knowing they were away that weekend. There she put the baby in their freezer. Afterward, she had returned to the house and faked the incident. The following day, once the police had finished searching her house, she had received a scare. Her parents called to say they were returning immediately. Paula had hastily retrieved the baby from her parents' freezer and kept it in her own basement freezer until the Wednesday, when she had made the two-mile drive to the riverside fishing area and dumped the frozen body in the trash can.

All this, of course, was mere conjecture. There was still nothing concrete to connect Paula Sims to the murder of her daughter, until detectives noticed something in her kitchen—a roll of black "Curb Side" garbage bags. They were manufactured by a company called Poly Tech Inc. for Kmart, and in make, size, and color they looked to be identical to the bag found with the body. Investigators wondered if there was any way of discovering if one of these bags had been used to hold Heather's body.

Bags of Evidence 77

Because kidnapping is a federal offense, the FBI had been closely involved from an early stage in the abduction of Heather Sims, and, in May, their agents visited Poly Tech in Minneapolis. What company Vice President Rex Warner first had to say threw a real damper on proceedings; each day, he explained, Poly Tech made in excess of 1 million bags for Kmart. While the agents tried hard not to show their disappointment, Warner explained that all was not lost. Identification was still a possibility. It all had to do with the manufacturing process.

Garbage bags are made from huge sheets of melted polythene resin that pass over rollers. In one procedure, a blade makes the perforations that enable the bags to be ripped from the end of the roll; elsewhere on the production line, a hot blade called a heat seal melts the bottom of the bag and seals it tightly.

Lab workers perform the early stages of mitochondrial DNA extraction at the FBI's state-of-the-art crime lab in Quantico, Virginia. *(AP Photo/Charles Dharapak)*

THE FBI LABORATORY

When J. Edgar Hoover became director of the Bureau of Investigation in 1924, he took a keen interest in the impact that science was having on crime solving. This impact mostly took the form of fingerprint identification, but as other scientific techniques became popular, Hoover encouraged the bureau to closely monitor developments. Much of their attention was centered on the work of Colonel Calvin Goddard, and in 1931 the bureau's top forensic agent, Charles Appel, was sent to Chicago to learn all that he could from Goddard. It was a tough learning curve. In the space of just two months, Appel was expected to pick up the rudiments of serology, toxicology, handwriting, and typewriter analysis. Appel brought these skills back to Washington, D.C., and in July 1932 he proposed that work should begin on a "criminological research laboratory."[3]

Four months later his wish came true: On November 24 the bureau's forensic science laboratory was officially opened in Room 802 in the Old Southern Railway Building on the corner of 13th Street and Pennsylvania Avenue. It was stocked with state-of-the-art equipment (for the time): a new ultraviolet light machine, chemicals, a microscope (borrowed, not bought), a helixometer, and some wiretapping apparatus. Over the next couple of years, a separate fingerprinting section was added, along with a photographic department.

With millions of feet of plastic processed daily, there are variations. For instance, the thickness of the sheets may vary slightly, or else imperfections in dye color can be found. But the most noticeable imperfections come from dust and fragments of plastic that build up on the rollers or the heat seal; these can leave identifiable marks on all the bags that run through. To minimize this, the heat seals are changed at least every 24 hours. What this means, though, is that while bags made within seconds of each other are microscopically similar, bags made at either end of the working day will display markedly different heat-seal characteristics.

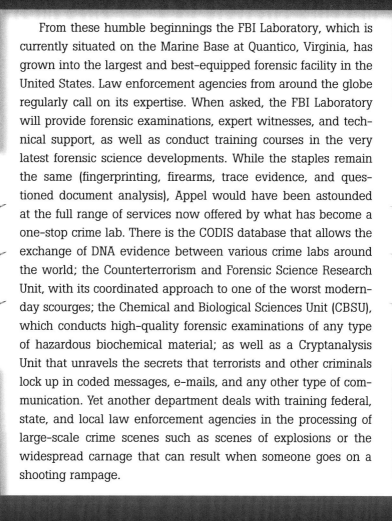

From these humble beginnings the FBI Laboratory, which is currently situated on the Marine Base at Quantico, Virginia, has grown into the largest and best-equipped forensic facility in the United States. Law enforcement agencies from around the globe regularly call on its expertise. When asked, the FBI Laboratory will provide forensic examinations, expert witnesses, and technical support, as well as conduct training courses in the very latest forensic science developments. While the staples remain the same (fingerprinting, firearms, trace evidence, and questioned document analysis), Appel would have been astounded at the full range of services now offered by what has become a one-stop crime lab. There is the CODIS database that allows the exchange of DNA evidence between various crime labs around the world; the Counterterrorism and Forensic Science Research Unit, with its coordinated approach to one of the worst modern-day scourges; the Chemical and Biological Sciences Unit (CBSU), which conducts high-quality forensic examinations of any type of hazardous biochemical material; as well as a Cryptanalysis Unit that unravels the secrets that terrorists and other criminals lock up in coded messages, e-mails, and any other type of communication. Yet another department deals with training federal, state, and local law enforcement agencies in the processing of large-scale crime scenes such as scenes of explosions or the widespread carnage that can result when someone goes on a shooting rampage.

Armed with this information, scientists at the FBI's Laboratory in Washington, D.C., began comparing bags found at the Sims residence with the murder bag. First, they needed to know that all the bags were made from the same plastic. The answer was provided by David Nichols, a chemist and biologist from the lab's material-analysis section. He had no doubts: All the bags he tested were manufactured from the same type of plastic.

David Attenburger was a document and tool-mark examiner. He studied the heat-seal marks on the murder bag, the bags in the opened roll of bags at the Sims's house, and two bags that the Sims had filled with trash. It was immediately apparent that Heather's bag had a slight puncture hole that been caused by roller debris and a tiny opening on the heat seal where the plastic had not melted thoroughly. Under the microscope, the bags from Paula's kitchen displayed identical flaws. Attenburger concluded that "the same heat seal was used to make those bags in a relatively short period of time."[4]

In order to narrow the time frame even further, Attenburger examined yet another characteristic that accrues when plastic is pulled over production line rollers. A very slight stretching occurs, leaving marks that are called die lines. When the die lines on Heather's bag were compared with the next bag on the roll in the kitchen—much like the way in which ballistics experts compare the striations on bullets—they matched exactly. All this allowed Attenburger to later testify that the bags in the Sims kitchen and the bag used for Heather were made "within ten seconds"[5] of each other. The bags remaining on the roll, he said, had the same characteristics as the others, but a few additional marks indicated that they had been manufactured some time after the others. Put another way, if 13 bags were made in 10 seconds, it meant that the kidnapper would have had to wrap Heather's body in one of only 13 bags in the world similar to those in the Sims household. The odds against that occurring randomly were almost incalculable.

The barrage of evidence from the FBI lab kept on coming. Although Allen Robillard, chief of the hair and fibers unit, concluded that none of the bags had been connected directly to each other, the unique markings in the plastic overwhelmingly suggested that the bag with the body had been the first one torn from the roll at the Sims house. This was direct evidence of the strongest kind, establishing a clear link between Paula Sims and the garbage barrel where her baby had been dumped.

It had taken three years and another tragic death, but eventually brilliant work from the FBI crime lab was able to bring Paula Sims to justice. On January 30, 1990, after a three-week trial, she was convicted of murdering Heather. Despite strenuous prosecution attempts, the jury declined to send Sims to death row, and she was sentenced to life

Bags of Evidence 81

Police officers lead convicted murderer Paula Sims away from an Illinois courtroom after a hearing. *(AP Photo/*Edwardsville Intelligencer, *Jeremy Paschall)*

in prison without parole. In a separate plea agreement relating to the death of Loralei, Sims pleaded guilty to certain charges in return for an assurance that she would escape another murder trial. Later, during an appeal, it was revealed that, in conversation with her lawyer, "Paula admitted that she killed her two baby girls."[6]

Time of Death

Crime laboratories ordinarily go to extraordinary lengths to achieve the best possible working conditions. In order to accomplish this goal, nature is constantly manipulated. Germs and bacteria are eliminated as much as possible; air-conditioning provides a reliable working temperature; everything is subject to the highest standards of sanitation. The scientist needs this because, when analyzing evidence or conducting experiments that might send someone to death row, any hint of contamination could be catastrophic.

But there is a world-renowned crime laboratory that throws all these conventions out the window. Within the confines of the Anthropological Research Facility at the University of Tennessee in Knoxville, nature is allowed to run riot (with only the occasional outside intervention). The world knows this place as the Body Farm, and its sole function is to learn what happens to the human body after death.

Founded in 1980 by Dr. William M. Bass and surrounded by barbed wire, the Body Farm covers 2.5 acres of land adjacent to the university football stadium. At any one time there are dozens of bodies (most have been willed to science) dotted about the wooded scrubland. They arrive in various states—some headless, some embalmed, some naked, many unidentified—and they are placed in a variety of locations: car trunks, under canvas or plastic, buried in shallow graves, covered with brush, or submerged in ponds. The newly dead may lie alongside piles of disintegrating bones. They are deliberately exposed to wide extremes of

Forensics trainees excavate and mark a shallow grave at the Body Farm at the University of Tennessee Forensic Anthropology Facility in Knoxville. *(Peter Menzel/Photo Researchers)*

temperature. Some might be refrigerated in total darkness, for example, while others are left in direct sunlight. And then they are studied.

When a person dies, the body starts to decay immediately as enzymes in the digestive system begin eating the tissue. Then the insects and the environment take over. Attracted by the smell of putrefaction, blowflies begin their ravages, followed by small rodents and other animals. In high temperatures, total decomposition can be startlingly fast. At the height of summer in a hot climate, a body can deteriorate from fully intact to bare bones in a mere two weeks.

Each stage of decomposition is recorded and analyzed, then added to the Body Farm's growing database, which is made available to law

(Continues on page 86)

A GRAVE MISTAKE

Grave robbing is a nasty business, but this looked to be a case of the robbers getting more than they bargained for. In December 1977 officers in Franklin, Tennessee, were called to investigate claims that someone had disturbed the grave of Civil War hero Colonel William Shy, who had been shot and killed during the Battle of Nashville on December 16, 1864.

Investigators made a gruesome discovery. A headless corpse was found in a squatting position on top of Shy's cast-iron casket. It looked as if someone had placed a recent murder victim in Shy's burial plot. What better place to hide a murder victim, one might say, than to bury it in an existing gravesite. When he was asked to investigate, Dr. William Bass certainly thought so. The corpse was a white male, an inch or two either side of 5 feet 10 inches tall, 175 pounds, and well dressed in what looked like a tuxedo. Judging from the flesh, which was still pink and in fairly good condition, Bass put the TSD at between two to six months. He noticed that there was a hole in the coffin. Bass took a flashlight and peered in. All that was left of Colonel Shy was a brownish goo, typical of what Bass would expect to see of a body that had been interred for over a century. As Bass continued his search of the gravesite he noticed two cigarette butts—a common find at crime scenes— probably left by the dead man's killers. Although unable to confirm the cause of death, Bass had no doubt that the man had met his death by foul means.

The police began sifting through the register of local missing persons. They came up blank. Then memories drifted back a couple of months, to the discovery of another headless corpse just outside Knoxville. Could a serial killer be at work? Bass compared reports on the two bodies and could see no similarities whatsoever, thus eliminating that line of inquiry.

Then came a breakthrough. Investigators had gone back to the grave and looked more thoroughly inside the coffin, where they found a skull. It now looked as if the killer had attempted to shove the body into Colonel Shy's coffin, only to wrench off the head in doing so. The shattered skull revealed the cause of death: a large-caliber gun fired at close range.

As soon as Bass saw the skull, doubts began to gnaw at him. The dead man's teeth were in awful shape, full of cavities, and without a single filling. Bass tried to reconcile the expensive clothes with the absence of dental work and came up short. At that moment the phone rang. A technician from the state crime lab sounded concerned. He had been asked to analyze fibers from the dead man's clothes and he was puzzled. Everything was natural; there was nothing synthetic anywhere. Nor could he find any labels. And the style of clothes looked oddly old-fashioned, with the trouser legs laced up the sides. Bass could feel his own face reddening by the second. Especially when the technician asked hesitantly, "Do you think this could actually be the body of Colonel Sly?"

"I'm starting to think that it is,"[1] came the reply. Far from dealing with a modern murder victim, Bass now realized that he was looking at a case of someone attempting to steal Colonel Shy's corpse and decapitating it in the process. He had been fooled by the heavy embalming procedure performed on the colonel's corpse. This had kept it in almost pristine condition until someone punched a hole in the hermetically sealed coffin and attempted to haul out the contents. The culprit was never caught, and Shy received another military burial.

The media seized on Bass's embarrassment, gloating over the fact that his original TSD was off by *113 years*. But Bass was smart enough to realize his own shortcomings and those of everyone else. The problem lay in a universal lack of knowledge about what happens to corpses after death. His solution was the Body Farm.

(Continued from page 83)

enforcement nationwide. The more precisely the researchers can measure decomposition in identifiable conditions, the more solid is their contribution to solving and prosecuting a crime.

Most of the Body Farm's criminal work is geared toward establishing what Bass calls time since death (TSD). This is often one of the most contentious and controversial areas of forensic science. Guilt or innocence in homicide trials often hinges on just one question—when did this victim die? Bass knows just how difficult providing the correct answer can be. This was especially true when he came face to face with one of the most loathsome killers of recent times.

In the fall of 1993 the marriage of Darrell and Annie Perry was teetering on the brink of extinction. Darrell, 26, had been in and out of jail more times than he could remember, while 23-year-old Annie had used her husband's absences to conduct a string of affairs. But all was not yet lost for the New Orleans couple. In November 1993 they retreated to a remote log cabin at Summit, Mississippi, a hamlet not far from Magnolia. It was one last attempt to sort out their problems. With them was their four-year-old daughter, Krystal. The family had been dropped off on November 6 by Darrell's stepfather, Alan Michael Rubenstein, who owned the cabin and used it as his weekend getaway home.

The next anyone heard from the Perrys was on November 16 when Darrell made a collect call to his stepfather and his mother, Doris, saying that the family was ready to come home. But when Rubenstein arrived to pick them up, there was no one home.

Twelve days later, he again drove up to the cabin, and knocked at the door. Still there was no answer. As he had forgotten his spare key, he was not able to check inside. He asked around to see if anyone had seen the Perry family and a neighbor recalled seeing Darrell taking off in a van with a couple of men who looked suspicious. Since Darrell was known to use crack cocaine, it was not an entirely unreasonable assumption that the suspicious men were involved in the drug trade in some way. Rubenstein drove back to New Orleans, not overly worried, just concerned. But as the days passed, the silence got louder.

Finally, on December 16, Rubenstein returned to the cabin, and this time he had his spare key. When he let himself in he was confronted by a terrible sight. Darrell and Annie lay on the living-room floor and Krystal was sprawled across a bed. Their bodies were hideously bloated, decomposing and crawling with maggots. Darrell and Annie had been stabbed repeatedly. Little Krystal had been strangled. The medical examiner estimated the killings had all taken place at around the same time—weeks, maybe months before.

It did not take long for investigators to uncover a suspect. Just days after the bodies were discovered, Mike Rubenstein filed a huge life insurance claim. The person insured was his four-year-old granddaughter, Krystal Perry, and the beneficiary was Rubenstein's wife, Doris. On April 27, 1994, Doris Rubenstein received a check for $200,063 (payout plus interest) from New York Life. Just days later she withdrew $200,000 and handed it over to her husband.

When detectives learned of this development they contacted New York Life in Louisiana and asked for a copy of the policy. It showed an application date of September 13, 1991. Buried deep in the fine print was a clause stating that the policy had a two-year waiting period for benefits. Just three months after the policy came into full force and the benefits became payable, Krystal was dead.

It sounded ominous and it smelled worse, but did it make Rubenstein a callous murderer? Although investigators could find no hard evidence to link him to the crime, they began digging deep into Rubenstein's background. What they discovered was truly shocking.

In 1979 Rubenstein had started a TV-related magazine and was looking to hire an assistant. The local employment office gave him the name of Harold Connor. Because Connor was inexperienced, and because Rubenstein had put up most of the money, he demanded that Connor take out a life insurance policy naming Rubenstein as the beneficiary. The policy was taken out in August 1979 with a sum insured of $240,000. Three months later Rubenstein invited Connor on a hunting trip with himself, 12-year-old Darrell, and two others. Connor's initial reluctance fell victim to Rubenstein's smooth tongue and he went along. It was his first-ever hunting trip, and it would be his last. When Michael Fournier, Rubenstein's crony and a recent parolee from federal prison, scrambled over a fallen log, his shotgun slipped and discharged. The blast hit Connor in the back, killing him in minutes. It sounded like a tragic hunting accident, the kind that happens all too often. Rubenstein, still grief-stricken, submitted his claim, only to learn that the policy had a two-year waiting period before the benefits would become payable. Mutual of New York was off the hook and Rubenstein was out of pocket. He decided to sue, claiming that the insurance company had deliberately withheld information from him. The suit did not go well for Rubenstein. In deciding for the insurance company, United States District Judge Charles Schwartz Jr. wrote that Connor "was killed under highly suspicious circumstances, circumstances that suggest something far more sinister than a mere 'accident.'"[2] Despite this, no charges were ever leveled against Rubenstein, who was left to lick his wounds and to make sure that, next time, he got the timing right.

At least this was the scenario sketched out by investigators looking into the cabin murders. The problem was that there was no direct evidence to implicate Rubenstein, and prosecutors knew just how reluctant Mississippi courts traditionally were to convict on circumstantial evidence alone. As a result, the deaths of the Perry family remained officially unsolved for several years. Then, on July 3, 1998, the mood in Mississippi courts abruptly changed. On that day, a Pike County jury convicted Glen Conley Jr. of drowning three-year-old Whitney Berry to claim a $200,000 insurance payout. He had taken out the policy just one month before the tragedy. Despite a lack of direct evidence, the jury had no qualms about convicting him of cold-blooded murder for money.

The Conley decision refocused interest on the Perry murders. The circumstantial evidence against Rubenstein was just as strong as it had been against Conley, but the passing years made it more difficult for a prosecution to make its case. Even so, on September 17, 1998, the Pike County grand jury indicted Rubenstein on three counts of murder. Confidence ran high that Rubenstein would be convicted, but question marks still haunted the inquiry. What investigators needed was an expert to say when the Perrys had died. Someone suggested contacting Bill Bass at the Body Farm.

Bass got the call in May 1999. With the bodies long since buried, he had to rely on crime scene photographs and contemporaneous notes to reach his conclusions. The photos were hideous and informative. All three victims showed advanced signs of decompositions; the bodies were bloated from the bacterial activity within; dark, greasy stains—caused by fatty acids being released during the tissue breakdown—haloed each body; and the hair was beginning to slough off.

This was exactly what Bass expected to find. Years of laboratory research had taught him that the process and order of decomposition is highly predictable. What varies—sometimes dramatically—is the rate at which decomposition occurs. The biggest factor in dictating this chronology is ambient temperature. Higher temperatures accelerate bacterial decomposition as a body putrefies. Insects, too, prefer to harvest when the weather is hot. What this means is that a body dumped by a steamy Louisiana bayou in early August will probably be skeletonized by September, while there is a good chance that someone buried in a Minnesota snowdrift in November might still be intact the following spring. It all depends on how cold it is.

In an attempt to quantify the temperature–decomposition ratio, Bass and his colleagues have developed a unit of measurement that they call "accumulated degree days" (ADDs). This works by averaging the daily temperatures. To give an example: 10 consecutive 70-degree days at the height of summer would amount to 700 ADDs. This same number would also be achieved by 20 consecutive 35-degree days in winter. Cross-referencing with records at the Body Farm would provide typical examples of the level of decomposition that would be expected for 700 ADDs.

In the case of the Perrys, Bass saw signs of advanced decomposition. Visual comparisons told him that putrefaction at this level was consistent with 800 ADDs. Now he needed to know what kind of temperatures this part of Mississippi had experienced during the weeks prior to the bodies being discovered. The relevant weather data told him that it had been unseasonably cool, with subzero overnight temperatures on no fewer than eight occasions. From this information, and by backtracking the ADDs, Bass was able to conclude that the Perrys had been killed somewhere between 25 to 35 days before they were found (or between November 11 and November 21).

But something was wrong. The level of insect depredation that Bass saw in the photographs did not fit with the decomposition time frame.

Blowflies are very efficient when it comes to disposing of human flesh. And they love blood. Just its smell will bring them swarming in the thousands. When they arrive, they lay eggs that hatch into maggots a matter of hours later. Bass studied the photos under a magnifying glass. Judging from the size of the maggots, which were fully mature, he concluded that they had hatched approximately 14 days earlier. Ordinarily Bass would have expected to see empty pupa casings—about the size of a grain of wild rice—all over the bodies. But, try as he might, he could not find a single one. This cast a seemingly unfathomable pall over Bass's TSD calculations. Whereas the decomposition suggested a murder date in the middle of November, the absence of the casings, allied

SEEDS OF DOUBT

When the strangled body of Louise Almodovar, a 24-year-old waitress, was found in Central Park on November 2, 1942, suspicion immediately centered on her estranged husband, Anibal. But he appeared to have an ironclad alibi. There were several witnesses who placed him at a dance hall at the time of the murder. It was only when detectives looked more closely that they realized that Almodovar could have sneaked out the back door of the club, murdered his wife, and returned without anyone being the wiser.

A major clue came when Alexander O. Gettler, head of the OCME's Chemical and Toxicological Laboratory, examined crime scene photos and noticed that the body was lying in an unusual type of grass. Enlargement of the photographs allowed him to identify the individual strain of grass. Coincidentally, grass seeds of the same type had been found in Almodovar's pockets and trouser cuffs, yet he insisted that he had not visited Central Park for over two years. Any seeds in his pockets, he said, must have been picked up on a recent visit to Tremont Park in the Bronx. He was wrong. Joseph J. Copeland, professor of botany and biology at City College, later testified that the grasses in question—*Plantago lanceolata*, to the size of the maggots, shifted this time forward to approximately two weeks before the bodies were discovered, or December 2. From December 2 onward, Rubenstein had a watertight alibi. Bass stuck to his original TSD, though, confident that cold weather had retarded the blowfly activity.

Bass knew he was in for a grilling when Rubenstein's trial began on June 1, 1999. As he suspected, the defense came at him with both barrels. Bolstering their confidence was the claim by Tonya Rubenstein—a niece of the defendant—that she had seen Annie and Darrell Perry alive and well in a local bar called Mudbugs on December 2, 1993. If true, Rubenstein was in the clear, courtesy of an unbreakable alibi. According

Panicum dichotomiflorum, Eleusine indica—were extremely rare and grew only at two spots on Long Island and three places in Westchester County. The only place in New York City where such grass occurred was Central Park. Moreover, it could be further isolated to the very hill where Louise's body had been found.

Almodovar panicked, suddenly recalling a walk he had taken in Central Park two months previously, in September. Copeland shook his head. The grass in question was a late bloomer, mid-October at the earliest, so Almodovar could not possibly have picked up the seeds in September. But they most certainly could have gotten into his pockets on November 1.

At this, Almodovar broke down and confessed. He had arranged to meet his wife in Central Park; they had quarreled again, and he had killed her in a fit of rage. Later, claiming that this confession had been extorted from him under pressure, he pleaded not guilty. But it did no good. When a sentence of death was passed, Almodovar, despite being shackled from head to toe, fought so fiercely that nine guards were needed to restrain him. On September 16, 1943, he died in the electric chair.

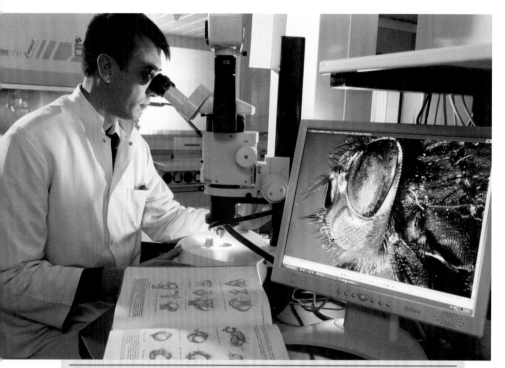

A forensic scientist uses a light microscope to identify a fly found on the corpse of a murder victim. A camera attached to the microscope displays the image on the computer screen at right. Identification of the insects on a corpse can provide vital information on time of death. *(Philippe Psaila/Photo Researchers)*

to the defense, the murders were less than two weeks old when the bodies had been discovered, thereby exonerating Rubenstein completely.

Although most members of the jury took the view that Rubenstein was a heartless killer, one holdout could not be swayed. Eventually, the jury announced it was hopelessly deadlocked at 11-1, and a mistrial was declared. The state would have to start all over again.

The hiatus gave Bass more time to think. Still puzzled by the blowfly/decomposition conundrum, he researched the crime scene more thoroughly, and learned that the cabin, far from being some rough-hewn shelter in the woods, was actually constructed from solid lumber. It was so well built, in fact, that Bass had little doubt that it would have kept the smells of death sealed inside, and the insects out for far longer than normal. Such a circumstance could retard blowfly activity by several days, pushing the TSD further back toward Bass's original estimate of approximately one month.

Opening arguments in Rubenstein's retrial began on January 27, 2000. As in the first trial, everything hinged on the time of death. Bass came under a strong attack from the defense as he outlined his new theory of how the cabin's solid construction had probably thrown off the blowfly calculations by several days, and he stood by his opinion that the Perrys had been murdered in mid-November, not early December as the defense claimed. After a bruising session on the stand, Bass remained in court to listen to the testimony of the medical examiner. Dr. Emily Ward used enlargements of the autopsy photos to show the jury the extent and horror of the injuries to the victims. Suddenly Bass sat bolt upright. These photos were new to him, and in a close-up of Krystal's face and head he saw a familiar brown object. And then another. And another. There could be no mistake—pupa casings.

Bass leaned excitedly over to the prosecutor and demanded to be put back on the stand. He told the jury that these pupa casings were scientific proof that blowflies had been feeding and laying eggs on the corpses for more than 14 days. When those earlier witness claimed to have seen the Perrys in a New Orleans bar on December 2, he said, they and their daughter were already dead and decomposing, more than 100 miles away. At a stroke, Bass demolished Rubenstein's alibi. And this time the jury only needed five hours to reach its decision.

On February 4, 2000, Rubenstein was convicted of first-degree murder. The next day he was sentenced to death. At the time of this writing he continues to appeal his conviction and sentence.

Ever since murder trials began, the "time since death" has been a source of constant friction and controversy. Nobody pretends that the Body Farm will provide all the answers to all the questions every time. But one thing is certain: Thanks to Dr. Bass and his extraordinary crime lab, the list of unanswered questions is getting shorter all the time.

A Valentine's Day Massacre

Arson, unpredictable and indiscriminate in the havoc it causes, is one of the most insidious of crimes. According to the National Fire Protection Association, in 2005 there were approximately 1,602,000 fires in the United States.[1] Of these, an estimated 323,900 were intentional. In human terms the cost is appalling—492 lives lost in 2005—while the monetary value of property damage exceeded $1.1 billion.[2] Setting aside wanton vandalism—half of all arson incidents are caused by juveniles—and those incidents triggered by mental disorder, arson usually has one of three motives: insurance fraud; revenge; or concealment of another crime, such as robbery or murder.

Fire has always exerted a powerful appeal for the criminal anxious to cover up evidence of illegal activity. The assumption is that a raging blaze consumes all, but this is rarely true. The murderer, for instance, anxious to dispose of his victim, will find that a human corpse is astonishingly fire-resistant, and that the kinds of temperatures essential for total destruction of teeth and bone are rarely encountered outside of a crematorium.

Another consideration for the criminally disposed to bear in mind is that a suspicious fire receives scrupulous investigation. Generally, at least three independent agencies are involved: fire officers, specially trained to locate the seat and cause of the blaze; the police, whose interest lies in discovering the perpetrators of any crime; and insurance adjusters, who are naturally eager to protect the interests of their employers. In

A fire research engineer from the ATF demonstrates a heat-release rate test with a cone calorimeter. Heat-release rate is a measure of the rate at which heat energy is emitted from a substance when burned. *(Getty Images)*

the most serious cases, investigative responsibility passes to the Bureau of Alcohol, Tobacco, Firearms, and Explosives (ATF). This government agency is unrivaled when it comes to investigating arson. From humble beginnings in 1886—two scientists working in the attic of the Treasury building in Washington, D.C.—the ATF laboratories now employ a staff of more than 100, working in four separate facilities. At the hub of this forensic empire is the National Laboratory Center in Maryland, which includes the Fire Research Laboratory (FRL), the first facility in the world dedicated to fire scene investigations. Agents from the ATF have seen it all, but the malevolence of the arsonist is something that no one ever gets used to.

THE FIRE RESEARCH LABORATORY

In 2003 the ATF opened its state-of-the art $135 million National Laboratory Center in Beltsville, Maryland. Standing on a 35-acre site, the 176,000-square-foot facility actually houses three separate laboratories: the Alcohol and Tobacco lab, the ATF Forensic Science lab, and the FRL. The latter is the first of its kind in the world. Here, scientists can analyze fires in an effort to answer, principally, two questions: How did this fire begin? and How did it spread? The conclusions they reach are used in criminal investigations, to train fire scene investigators, and to develop scientifically sound investigative methods for any law enforcement agency that deals with a fire.

The FRL's largest research bay is enormous, approximately 130 feet by 130 feet by 55 feet high (big enough to construct, for example, a small office building—indoors). Then the fire research engineers come in and torch the interior structure in a variety of different ways and in a variety of different locations. In this controlled environment, they can not only measure the heat-release rates of burning materials, but also replicate most of the fire scenarios encountered by fire investigators in the field. Automobiles, buses, even train cars—anything that can fit through the door—can be tested in the "burn room." Investigators are particularly interested in learning more about the often baffling path that smoke takes as it moves through a burning building; why someone may

At just after midnight on February 14, 1995, the Pittsburgh Fire Department responded to a five-alarm fire at 8361 Bricelyn Street in the Brushton district of the city. The house was home to the Buckner family, Ronald and Darlene, and their four children. None of the family was in the house when the fire department arrived. Of the many firefighters who rushed bravely into the blazing building, three never left alive. They were victims of a terrible misunderstanding. The house was built on a steep hillside, and from Bricelyn Street it looked to be a two-story die in one room, only for a person in the next room to escape unharmed.

Above the burn room looms a 60-foot-wide soot-blackened exhaust hood. As the smoke is sucked up, it passes through ducts where a calorimeter, capable of making more than 1,900 simultaneous measurements, records the volume and composition of the smoke. Scientists can track the intensity of the blaze and the release of toxic gasses, along with a host of other data. The environmental impact is kept to a minimum. All smoke is filtered and all toxins are removed before being released into the air.

Elsewhere in the building are smaller labs devoted to analysis of samples from the facility's research bays and from outside agencies. Each year the ATF takes part in about 1,500 fire investigations, up to 40 percent of which develop into full-blown arson investigations. Agents can often work a single case for many years. David Sheppard, a senior fire research engineer, sums up the lab's ethos this way: "We want to give investigators a leg up and try to give them an advantage. First, did a crime occur, and then try to figure out who did it."[3] Sheppard routinely catalogues his results and enters them on a database that can be accessed by law enforcement agents across the nation. Perhaps the most important thing Sheppard has learned in all his years of studying thousands of fires is this: "When the smoke detector goes off, get out quick."[4]

residence. Only from the rear was it apparent that there were four levels. When firefighters entered, they headed to what they thought was the basement to find the source of the fire. In reality, they had descended to the floor above the flaming basement, where smoke from the fire rose to surround them in blackness. When a fireball engulfed a stairway, it collapsed, plunging Captain Thomas Brooks, 42, Patricia Conroy, 43, and Marc Kolenda, 27, into the basement. There, fumbling in the dark, with their water hose burnt in half, they fought until their oxygen tanks ran out. Conroy was the first Pittsburgh woman firefighter to die in the line of duty. Another three firefighters were grievously injured. It was the blackest day to befall the Pittsburgh Fire Department since January 21, 1924, when seven firefighters had died while tackling an oil refinery blaze. Because of the fatalities, ATF was called in to investigate the blaze.

In any fire the following equation applies: fuel + oxygen + heat source = combustion. ATF agents needed to determine if any of these three ingredients was present in suspicious quantities. The house was built on four levels, with most of the damage occurring in the basement and attic. It appeared as if the fire started in the basement. Once it had consumed all the available oxygen and combustible material, it had then raced to the top of the house through gaps between the walls.

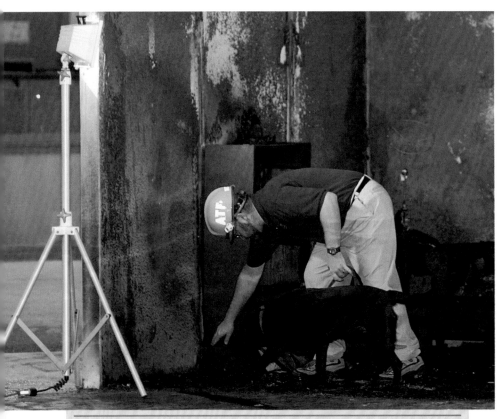

A K-9 trainer from the ATF guides his dog for a demonstrative investigation during a media tour of the FRL. *(Getty Images)*

As there was no obvious electrical cause for the fire—no frayed wiring or burnt-out appliances—attention centered on a pile of charred laundry that sat in the middle of the water-soaked concrete basement floor. Directly above this pile, the basement ceiling joists were burnt out in a circular pattern about 12 feet in diameter. Such a distinctive burn pattern immediately aroused suspicions.

All burning materials produce energy at varying rates. Using a table of known values for burning textiles, ATF agents calculated that a pile of laundry of that size and type of fabric would generate 176 kilowatts of energy. Another mathematical formula revealed the maximum height that flames from such a blaze would reach. In this case, it was just 30 inches. This left a gap of several feet to the ceiling. There was clearly something wrong. Even if the clothes had been stacked right up to the ceiling, they could still not have generated enough heat to cause such damage to the joists.

However, add a single gallon of gasoline to the laundry pile and suddenly there would be 1300 kilowatts of energy at work, powerful enough to project a flame 13 feet in the air, right to where the charred joists showed the trademark "alligatoring effect" burns that are typical of gasoline-induced flames. On February 16 the first samples from the fire were sent to the ATF National Laboratory in Rockville, Maryland, to be examined for the presence of accelerants.

While ATF investigators continued to unravel the mysteries of the increasingly suspicious fire, detectives began delving into the Buckner family background. They learned that the Buckners had rented the house on Bricelyn Street since November 1990, but that for several years they had been trying to buy their own property, only to be rebuffed for credit reasons. The latest rejection had come in the summer of 1994, when the deal fell through because the family lacked the necessary deposit. Despite being out of work and heavily in debt, suddenly, in November 1994, after four years of living without any kind of insurance, Darlene Buckner, 38, took out a renters' policy on 8361 Bricelyn Street for $20,000 from the Keystone Insurance Company. On the day of the blaze, she filed a claim and was given an immediate payment of $2,500. One week later she was pressing Keystone to advance her another partial payment. On February 23, Keystone cut a check for $1,500 and the next day Buckner placed a $500 deposit on the very house that had so taken her fancy the previous summer, 521 Grove Road, Penn Hill. Life for the Buckners had taken a turn for the better.

However, one week later, scientists at the ATF National Laboratory confirmed what fire scene experts had suspected from day one—this was no accidental blaze. There is a widely held belief among arsonists that most, if not all, accelerants are consumed in a fire, or else evaporate. This is a gross misconception. Traces invariably remain, and to find these, the investigator will once again direct his or her attention to the lowest point of ignition. Also, any accelerant used to start the fire has a tendency to be absorbed by the charred wood or seep into cracks in the flooring, where it often fails to burn from lack of oxygen. Good quality control samples—the kind that scientists prefer—are best harvested from these low locations and from any porous surfaces such as carpet and furniture.

On this occasion ATF scientists analyzed samples of laundry, wood, and cement chips from the burnt-out house by heating them with charcoal-coated strips, which capture any hydrocarbon molecules—the main component of accelerants such as gasoline. Next, the charcoal strips were placed in a pyrolysis gas chromatograph.

There are many different types of chromatography, but all work on the same basic principle. A substance is first dissolved in an appropriate solvent (this is called the mobile phase), then the sample is passed through an adsorbent (a solid, such as charcoal, that has the power to absorb large quantities of gases). This is known as the stationary phase. As the sample migrates, its molecules become the subject of a struggle between the stationary phase and the mobile phase, which concludes with elements in the sample attaching themselves to the adsorbent. Because each compound has its own distinctive adsorption rate, this adherence takes place at different times, and from these figures it is possible to identify the compound.

In pyrolysis gas chromatography—which is especially useful in identifying accelerants used in arson cases—the sample is heated to a temperature at which it decomposes into gaseous components. The scientist then compares the resulting chemical profile with a set of known reference values, and from this is able to identify the substance. When the ATF scientists finished running their tests on evidence from the Bricelyn house, they found that no fewer than six samples showed obvious traces of accelerant. This was now a clear case of triple murder.

Such a finding put the Buckners under an even more powerful microscope. The focus at first centered on the Buckners' teenage daughter, who had a history of problems involving fire, but she was quickly eliminated as a suspect. The investigative spotlight then focused on Darlene Buckner and her 17-year-old son from her first marriage, Gregory Brown.

On the night of the fire, Buckner said that she and Brown had driven in her Dodge Dynasty to a nearby Giant Eagle store to buy food for the upcoming wake of a recently deceased relative. They had left home, she said, at a few minutes past midnight. When they returned it was 1:05 A.M. and the house was ablaze. But a neighbor, Keith Wright, recalled seeing Brown standing in front of the Buckner residence, with smoke billowing from the rear basement window, prior to 12:22 A.M., when the first emergency vehicles arrived. Also, a police officer happened to talk to Buckner as she returned home. He distinctly remembered that she was alone in her car, at a time when she claimed that Brown was accompanying her. As the holes began to appear in Buckner's story, investigators delved into the day's events and worked up the following scenario.

Some time on February 13, Buckner removed a crate containing a one-gallon gasoline can from the trunk of her car and placed it on the front porch of her residence, giving as her reason that she had to transport relatives to the doctor the next day. She claimed that, at the same time, she had also removed a baby seat and placed it inside the house. Later that night, Buckner had left for the store at the time she claimed, but with one significant difference to her story—she was alone. Brown, investigators theorized, had remained behind to set the blaze. At approximately 1:05 A.M., Buckner returned alone—not with Brown as she claimed—to find the street full of emergency vehicles. Flagged down by the police and forced to park some way off, she ran down Bricelyn Street toward her residence. Quite by chance, the police officer she met had glanced inside Buckner's car at this time and spotted a baby seat in the front of the vehicle. He had inspected the car thoroughly to ensure that a baby had not been left behind. This was the same car seat Buckner claimed to have removed from the vehicle earlier in the day.

(Continues on page 104)

A STORY GOES UP IN FLAMES

On November 25, 1929, a car traveling near the Bavarian village of Etterzhausen crashed and burst into flames, incinerating the driver, Leipzig businessman Erich Tetzner. The next day his widow identified the blackened remains. She also filed insurance claims on Tetzner's life, totaling 145,000DM ($28,000), a huge amount for someone of Tetzner's relatively modest means. Alert to the possibility of suicide—the policies had only been in effect a few weeks—cautious claims adjusters told Frau Tetzner that an autopsy would be necessary.

When Germany's leading criminalist Richard Kockel, of the Institute of Forensic Medicine at the University of Leipzig, inspected the corpse, several features disturbed him. Inexplicably, the cranium and both legs below the thigh were missing; neither, Kockel thought, was likely to have been consumed in the fire. Also, the 26-year-old Tetzner had been described as burly and muscular, yet these remains were those of a slim, light-framed male. When Kockel found that the respiratory passages were entirely free from the kind of sooty deposits customarily associated with smoke inhalation, his suspicions went into overdrive.

Most fire deaths are caused not by flames or heat, but from the effects of carbon monoxide. The blood turns a bright cherry color and may retain detectable amounts of the lethal gas for months after death; but when Kockel tested blood samples, none showed carbon monoxide contamination. This meant that the victim was already dead when the car burst into flames. Eliminating stroke or coronary failure as being unlikely in someone so young, Kockel considered two alternatives: trauma sustained in the accident, or foul play.

Although gutted by fire, the car was still intact and did not appear to have sustained enough damage to instantly kill its driver. Kockel now pondered the missing cranium, wondering if it had been removed before the fire because it showed telltale signs of physical distress, a suspicion strengthened by the discovery of part of the dead man's brain on the road, almost two yards from the driver's seat.

Kockel was now convinced that the dead man was an innocent bystander, murdered as part of an insurance swindle, and that Tetzner was still alive and in hiding, and it would only be a matter of time before he contacted his wife. A phone tap led the police to a call box in Strasbourg, France. On December 4 a bulky man in his mid-20s was arrested in mid-call. So astonished was Erich Tetzner that he immediately admitted his true identity.

That night he made the first of numerous confessions. He had been planning the fraud for some time. His first murder attempt failed when the intended victim, a man of similar build to himself, managed to fend off the wrench-wielding Tetzner. Eventually, he picked up a hitchhiker, a wiry man about 20 years old. Tetzner had waited until the hitchhiker fell asleep and then doused his car with gasoline and tossed in a match.

Not true, said Kockel; the victim was already dead by that time. Moreover, why, if he had fallen asleep on the passenger side, was he found in the driver's seat? Caught in an obvious lie, Tetzner abruptly changed his tale. His next version had him accidentally running down the hitchhiker in the dark. Only after picking up the stranger and placing him in the car did he realize that he was dead. Suddenly an idea occurred to him and he perpetrated the swindle.

Again Kockel shook his head. In that case, why remove the missing body parts? It was one question too many. On May 2, 1931, Tetzner was executed at Regensburg.

Although rarely used as a means of homicide, fire has obvious attractions for someone seeking to destroy an unwanted body. However, recent advances in medical technology—it is now possible to ascertain whether burns were contracted ante- or postmortem—virtually guarantee detection.

(Continued from page 101)

The disaster on Bricelyn Street was a maddening crime. There was plenty of circumstantial evidence to suggest that Buckner and Brown had engineered a lethal insurance fraud, but nothing to tie them conclusively to the crime. Under questioning, Buckner's story began to waver—she suddenly shifted the time of her departure from Bricelyn Street to 11:45 P.M., 20 minutes earlier than previously stated—but she did not crumble. Nor did she buckle when, on March 15, her son got himself arrested on unrelated narcotics and firearms violations. Instead, Buckner continued her real estate negotiations. However, 10 days later, the owner of the Grove Road property decided to pull out of the deal and returned the $500 deposit. Buckner was not fazed. She immediately located another property and began negotiating its purchase, and she continued to hound the insurance company. On May 30 her persistence paid off when she received a check for $16,000 in full settlement of her claim. By coincidence, that same day, Brown was sentenced to the Vision Quest Boot Camp at South Mountain, Pennsylvania, for 90 days. It was here that the seeds of his downfall were sown.

Brown just could not keep his mouth shut. He bragged repeatedly to fellow inmates that he had deliberately set the fire so his family could claim the $20,000 insurance payout. There was no remorse for the "fireheads"[5] who had died in the blaze, just a jeering account of how "he stayed around the area . . . to watch the fire."[6] As it happened, details of Brown's verbal indiscretions stayed within the penal system, and there they remained, even after the teenager was released. He was immediately dispatched by his mother to Billings, Montana, to live with relatives (she thought that a change of environment might straighten out his errant ways). All the while, frustration gnawed away at the ATF investigators. Finally, on October 13, 1995, the agency offered a reward of $15,000 for information leading to an arrest and conviction in the Bricelyn Street murders. Only at this point did word of Brown's boot camp confession filter through to investigators. On April 12, 1996, Brown was arrested in Billings. That same day, Buckner was arrested at her new home in Pittsburgh. Mother and son were charged with homicide, arson, conspiracy, and insurance fraud.

Their trial commenced on January 31, 1997. The prosecution was dealt a stinging blow right at the outset when Judge David S. Sercone ruled that statements made in jail by Brown regarding his mother's involvement were not admissible. This greatly weakened the case against Buckner and placed an even greater burden on ATF case agent William J. Petraitis to ensure that justice was done. In more than six hours on the stand, Petraitis explained how he had tracked the fire's lethal journey through the house. It had burned a hole through the kitchen floor. This, and the mushroom pattern of burns on the joists, made it evident that the fire started in the basement. From there, it had climbed through the frame of the house, filling the family room with heat and smoke, but leaving it largely intact. Because of the unusually high temperatures involved, there could be no doubt that this fire had been set deliberately. When defense attorney Sidney Sokolsky contended that there had been no crime, that the fire had been a dreadful accident, maybe caused by a faulty fixture in the kitchen, Patraitis answered him coldly: "That theory doesn't work in physics."[7]

For the prosecution, led by Assistant United States Attorney Shaun E. Sweeney, it was simple. An attempted insurance scam had gone hideously wrong and left three firefighters dead. The jury agreed. On February 21, 1997, they convicted Brown on all charges and he was later given three consecutive life terms. His mother was treated more leniently. Acquitted of murder, Buckner was found guilty of insurance fraud alone, for which she received three years of probation, 500 hours of community service, and a $5,000 fine.

In 1998 the ATF deployed its first mobile laboratory, which was designed to allow examination of evidence at the scene of a fire or explosion. Such developments have kept the ATF at the forefront of fire investigation, reinforcing its position as the premier arson-fighting organization in the world.

Prints and Pixels

There are times when forensic science is a victim of its own success. As the technological advances come hurtling by, one after another, the courts sometimes struggle to keep pace. For most of the 20th century, science-based testimony in America was governed by *Frye v. U.S. 293 F. 1013 (D.C. Cir. 1923)*. In this landmark case—which effectively banished the polygraph or "lie detector" from the American courtroom—the courts decided that scientific evidence could only be introduced if the underlying principle was generally accepted by the scientific community. For seven decades *Frye* ruled the land and for the most part it worked very well. But concerns about the gathering pace of ethnological innovation forced the U.S. Supreme Court to reexamine the whole issue of scientific testimony. The results of their deliberations were rendered in *Daubert et al. v. Merrell Dow Pharmaceuticals, 509 U.S. 579, (1993)*. This shifted responsibility for assessing the admissibility of scientific evidence from the scientists to the trial judge. Henceforth, he or she would be the "gatekeeper" of what was acceptable in terms of scientific evidence. Many argued that this placed an unfair onus on the judge; after all, an in-depth knowledge of law is no guarantee of scientific competence. This skepticism prompted some states not to adopt the *Daubert* guidelines. Instead, they kept faith with the tried and tested *Frye* "general acceptance" decision. One of these holdouts was Washington State. Its supreme court saw no reason to tamper with something that had served the legal system so well for so long. Not long after reaching this conclusion, the Washington judicial system found itself tested by one of the most contentious forensic scientific innovations to come along in recent times.

On the evening of Saturday, May 13, 1995, a resident at the Salish Village condo development in Kirkland, Washington, noticed that a neighbor's door was ajar. It was still open the next morning. At 9:40 A.M. the now greatly concerned resident knocked tentatively and then entered. On the floor of the bedroom lay the cold, lifeless body of 27-year-old Dawn Fehring. The top bed sheet and a T-shirt had been wrapped around her head and neck. Apart from this she was naked.

Neighbors were stunned. Salish Village was a quiet place and Dawn was an exceptionally reserved young woman. Just three weeks earlier she had returned from Japan, where she had taught English for two years, and enrolled at the Lutheran Bible Institute in Issaquah, some 16 miles away. She was house-sitting the condo for friends who were still in Japan as missionaries. Dawn, too, hoped to follow this path. She had no boyfriend, no one who bore her a grudge. Nonetheless, someone had killed her.

The apartment showed no signs of a break-in, nor could friends and relatives imagine someone as cautious as Dawn admitting a stranger. So how had the killer gained admittance? The police were forced to speculate. They knew that Dawn had been scheduled to attend a Mother's Day party with her family on Sunday. If she had been baking food for the event—a family tradition—it was possible that she might have left the front door open for ventilation, allowing the killer to sneak in undetected.

She had last been seen alive the previous Friday, by a clerk at a nearby chain store who happened to notice Dawn shopping. This was significant as, according to the medical examiner, she was probably murdered that same day. The autopsy gave the cause of death as "mechanical asphyxiation."[1] It appeared as if the killer had pressed either a pillow or bed sheet over Dawn's face and not let go until she was dead.

Despite clear evidence of a particularly violent sexual assault that left Dawn battered, there were no DNA traces from her attacker. In fact, the crime scene was remarkably clean, apart from some cigarette ash found on a bedside table. Since Dawn was a nonsmoker, it was reasonable to assume that this came from a cigarette smoked by the killer. The only other clue was a pale pink bed sheet stained with bloody smudges (as if the killer had wiped his hands) and a few blood spots on the carpet. All the blood was tested and found to have come from Dawn.

With no obvious suspects from Dawn's circle of friends, detectives began questioning other residents of Salish Village. One of those they spoke to was Eric Hayden, a 32-year-old mill worker who lived one flight up from Dawn, on the opposite side of the stairwell. When asked to account for his whereabouts on the Friday evening, he seemed edgy and instinctively grabbed a cigarette before answering. His recollection was hazy, and consisted of a claim to have been out drinking with some friends. But when asked to provide names, he drew a blank. When detectives spoke to Hayden's girlfriend, she said that she had been away that weekend, but later Hayden had told her that he had been too drunk to remember where he had been on the night of the murder.

Gut instinct told the detectives that there was something suspicious about Hayden's demeanor and his story. Despite the detective's suspicions, being drunk to the point of amnesia hardly amounted to proof of murder, and at first glance there was nothing to connect Hayden with the murder. When checks were done on Hayden's background, it became clear that he did have a history of problems with alcohol. In January 1995 he had been taken into custody on a drunk-driving charge. When he had failed to show up for a hearing, he was arrested and his fingerprints were taken. Those records might prove useful, detectives thought, especially if their suspicions about the bloodstained bed sheet proved to be accurate.

Everyone who examined the sheet agreed that it did look as if smudged fingerprints were visible in the bloodstains. But everyone in law enforcement knew that lifting fingerprints off any kind of fabric was next to impossible. For this procedure to even have a chance of success, the weave of the cloth must be extremely tight. If it is an open weave, all definition is lost. Here, the bed sheet *was* made from close-woven cloth, but it was still beyond the bounds of conventional fingerprint analysis. The fitted bed sheet was passed to Daniel Holshue, a King County latent print examiner with plenty of experience in difficult cases. He examined every inch of the sheet, and then cut out the five areas that held the most blood and smudged prints. Next, he treated these samples with amido black, a dye stain that reacts with organic proteins such as blood. At every step of the way he was haunted by one overriding fear—what if such a drastic technique destroyed the only piece of genuine evidence? At first his anxiety seemed justified as the whole sample turned navy blue, but rinsing in pure methanol gradually lightened the background, to the point where protein-based handprints became faintly visible. As a final step, Holshue dipped the samples in distilled water to set the prints. Then photographs were made of the results.

After all these chemical processes were completed, Holshue took a 5x magnifying glass and studied the results. He was bitterly disappointed. The contrast between the latent prints and the pieces of bed sheet was still too subtle for Holshue to isolate the minimum eight points of comparison legally required in Washington State for a positive identification. It was like peering into a maze, as the weave of the fabric and the whorls and ridges of the prints themselves merged into one swirling whole.

At this point, Holshue realized that he had run out of technology, but he still was not defeated or dismayed. He knew there were experts in other fields that he could call on. One of these was Erik Berg, an imaging expert at the Tacoma Police Department, which had been using forensic digital enhancement since January 1995. Berg had spent the previous two years developing a software package that would digitally enhance crime scene photographs.

Open any modern magazine and chances are that it will contain digitally enhanced or restored photographs. This is especially true in the advertising sections, where photo editors dodge, burn, blur, sharpen, and airbrush their way to the desired effect. Image enhancement is not new. It was developed by jet propulsion labs in the NASA space programs in the 1960s, where scientists were eager to isolate galaxies in images of outer space. The process is expensive and time-consuming, two reasons why most law enforcement agencies were reluctant to embrace it. There is also the tricky issue of authenticity. The whole point of digital enhancement is to alter an existing image. When done forensically, a strong case can be made for saying that this

(Continues on page 112)

HACKING THE BAD GUYS

The presence of computers in every facet of modern life means that computer forensics is now one of the fastest growing branches of forensic science. Certain specialized labs deal with this discipline and no other. There are many reasons to call on their services, including the following:

1. To analyze computer systems belonging to defendants (in criminal cases) or litigants (in civil cases)
2. To analyze a computer system after a break-in; for example, to determine how the hacker gained access and what the hacker did
3. To gather evidence against an employee that an organization wishes to terminate

The following example deals with gathering data for use in a criminal prosecution. At every stage in the process, the investigator needs to tread lightly for fear of stepping outside the law. Say, for example, a warrant is issued to examine a suspect's laptop. Even if the investigator subsequently visits the suspect's home and finds a stack of desktop computers, none of these can legally be examined without the issuance of a new warrant. For this reason, strict codes of conduct have evolved. The following is standard procedural practice when dealing with any computer thought to contain useful evidence:

1. Secure the computer system to ensure that the equipment and data are safe. The onus is on the investigator to ensure that no unauthorized individual can access the computers or storage devices involved in the search. If the computer connects to the Internet, this connection must be severed.
2. Make a reference image of the hard disk. This is essential to counter possible future accusations of disk tampering.

3. Make a copy of all the files on the system. This includes files on the computer's hard drive or in other storage devices. Since accessing a file can alter it, it is vital that investigators work only from copies of files while searching for evidence. The original system should remain preserved and intact.
4. Find every file on the computer system, including files that are encrypted, password protected, hidden, or deleted but not yet overwritten.
5. Recover as much deleted information as possible using applications designed to detect and retrieve deleted or hidden data.
6. Attempt to decrypt and access protected files.
7. Analyze any unallocated space on the hard disk. This is an area of the drive that is normally inaccessible. It might contain files or parts of files relevant to the case.
8. Document every step of the procedure. This cannot be stressed strongly enough. The computer scientist must be able to demonstrate to the court's satisfaction that the investigation preserved every byte of information on the computer system. Without the proper documentation, the evidence may not be admissible.

Professional cybercriminals, of course, are aware of all these procedures, and most will take steps to cover their tracks. However, as many have learned to their cost, no matter how meticulously they try to wipe the contents of a hard disk, there is a software package, somewhere, that can defeat them. Experts reckon that just about the only way to guarantee that no one will be able to read the deleted contents of a hard disk is to grind the drive into tiny fragments, preferably dust, and then to dispose of that dust very carefully in several different locations.

(continues)

112 CRIME LAB

(continued)

A recent development in the cybercrime wars is the introduction of what are known as "anti-forensics" programs. As the name implies, their purpose is to scramble information on a hard disk (or any other data storage device), making it virtually impossible to retrieve any data. Some justify these programs with claims that they highlight the inherent flaws in gathering computer data. It is, they say, just too easy to manipulate timestamps and filenames—weaknesses that should render any such evidence unreliable in a court of law. While the legal battle rages on, it is undeniable that certain criminal elements use anti-forensics programs in an effort to thwart the crime lab.

(Continued from page 109)

amounted to tampering with evidence. Berg, acutely aware that any defense lawyer was sure to come after him with guns blazing, neatly sidestepped this possibility with a simple expedient. When he received the photographic images, his first course of action was to encrypt and store the originals, using standard government software, to negate any potential claim of photo tampering.

Berg next loaded the copy images onto his computer. To further ward off possible accusations of evidence tampering, he created an authentication system designed to protect the photo's pixels and to record every mouse click and keystroke he performed, so that any expert could repeat the process. Berg selected the best image, which showed an apparent palm print, and got to work. Using filters from a digital-enhancement program he had written himself, called More Hits, as well as filters from the software package Adobe Photoshop, he began by attempting to improve the sharpness and image contrast. At first he struggled to make any headway. It was still a poor-quality image, with the ridge patterns from the palm print and the fabric pattern intermingled. Berg turned to a collection of pattern and color isolation filters, designed to remove any unwanted colors and distracting elements. This allowed him to blur the background weave pattern. The trade-off came in image contrast. To overcome this drawback, Berg employed another filter, one that raised the tonal values of the ridge details, much like increasing the contrast on a television set. Finally, he was looking at what he thought was a useable print. He repeated the process with all the images and then sent them back to the King County Police Crime Laboratory.

Holshue was astonished by what Berg had achieved. On June 5 he ran the prints through an automated fingerprint checking system and it generated a match almost immediately. Holshue then carried out a manual check. There could be no doubt about it. On one of the fingerprints he located 12 points of comparison with the criminal record on file; and when it came to the palm print the comparison number jumped to 40, incontestable proof of identification. Holshue reached for the phone. As he put it later: "It's always great to call up a detective and say, 'he's your guy.' . . . It's really satisfying."[2]

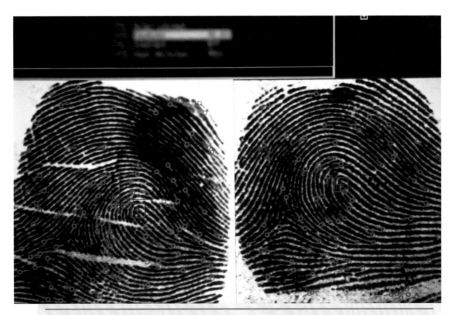

In a screenshot of fingerprints being compared using a computer program, the purple tags identify a set of features for each fingerprint that, in combination, are unique enough to enable the fingerprints to be matched. *(Patrick Landmann/Photo Researchers)*

114 CRIME LAB

Two hours later Eric Hayden was charged with murder.

The crime lab had pulled off a forensic miracle, but would the courts admit Berg's evidence? This was by no means certain. When Hayden's trial opened on December 26, 1995, his defense counsel, Andrew Dimmock, claiming that digital enhancement was an unproven technique, argued to have the prints evidence thrown out. The outcome was a

A GRAPHIC CRIME

The often-volatile relationship between San Francisco business brothers Jim and Artie Mitchell ended on the night of January 27, 1991, when Artie's girlfriend Julie Bajo called 911 to report a shooting at Artie's home. While she was on the line more shots rang out in the background. When the police arrived they found Jim Mitchell, half-dazed, carrying a .22 rifle and a holstered handgun. His brother's bullet-riddled body lay on the floor. Jim admitted shooting Artie, but he claimed it had resulted from an unpremeditated loss of temper. Skeptical prosecutors charged him with first-degree homicide.

The 911 tape was given to Dr. Harry Hollien, an expert in forensic acoustics. He studied it, attempting to isolate and identify where each shot had been fired. The first three shots were fired in 25 seconds; then came a half-minute gap before shots four and five were fired in quick succession. This gap was pivotal to the state's contention that Mitchell had waited for some time and then deliberately gunned down his already wounded brother. The major difference between murder and manslaughter is one of intent; if Hollien's interpretation was correct, then it would seem to undermine the defendant's claim that the shooting had happened in the heat of the moment.

At trial the prosecution also submitted a computer-generated 3D video animation of the shooting. It showed Artie approaching his bedroom door, walking down the hall, where

"*Frye* hearing," to determine the admissibility of fingerprints identified by use of the enhanced digital-imaging process.

The state presented testimony from Holshue and Berg, who explained the steps they took to ultimately identify Hayden's palm and fingerprint from the fitted bed sheet. Holshue had little difficulty getting the court to accept his use of the amido black chemical dipping

> he was shot twice, then entering the hallway and getting shot the final time in the head. The shots were represented by red laser-like lines against a blue background. Not only did this tape claim to reproduce the crime scene, it also purported to show the order in which the shots were fired. Clearly, this was moving into the realm of speculation.
>
> Criminalist Lucien Haag, whose crime scene reconstruction had inspired the video, explained the methodology behind his conclusions. He had traced the bullet paths through the various rooms, noting angles and impact points. Because one bullet had passed through a door, he purchased a door of the same type, just to measure what level of deflection it might exert. Under cross examination Haag admitted that the video was a subjective interpretation and conceded that with so many shots there were potentially thousands of possible sequences to consider.
>
> This was a point hammered home by the defense. Crime scene expert Charles Morton felt that the whole subject of reconstructions was a forensic quagmire. He argued that the criminalist's role is to analyze the physical evidence available and present any conclusions to the court in a fair and impartial manner. Thereafter, it is up to the jury.
>
> On February 18, 1992, Mitchell was acquitted of murder, but convicted of manslaughter and imprisoned for six years. He died in 2007 of an apparent heart attack.
>
> Since this case, 3D reconstructions have figured prominently in several other trials.

process. There was plenty of precedent for such a procedure. Berg had the more difficult task. He admitted that as early as 1987 the FBI had expressed concerns about the use of digital enhancement in fingerprint identification, citing worries "that nonexistent detail is added to the latent print."[3] This skepticism, Berg admitted, had retarded research in the field, but he asserted that many of the objections had now been overcome. He was able to demonstrate that, as used by himself and others, digital enhancement is a subtractive process in which elements are removed or reduced; nothing is added. The software he used prevented him from adding to, changing, or destroying the original image. In contrast with "image restoration," a process whereby details not present are added to achieve the desired end result, image enhancement merely makes what is there clearer and more usable. He stressed that there is no subjectivity in the process—if the detail is not there, he cannot add it. The technique, he said, has a reliability factor of 100 percent and a zero percent margin of error. His results were easily verifiable and could be easily duplicated by another expert using his or her own digital camera and appropriate computer software.

To further strengthen his case, Berg set up his equipment in court and demonstrated how he had obtained his results. Digital photographs work with light sensitivity, just like film photographs, except the computer uses a chip and a hard drive in place of the camera's film. He explained that digital photography's big advantage over its analog film counterpart is that it can capture no fewer than 16 million colors and differentiate between 256 shades of gray. Such wide-ranging parameters provide almost infinite scope for the color adjustment necessary in fingerprint identification. To back up Berg's claims, the state also introduced literature showing that the Los Angeles County Sheriff's Department had been using digital image processing as a means of enhancing latent fingerprints since 1987.

By contrast, the defense produced no witnesses and no literature to refute Berg, instead mustering just a half-hearted plea for his testimony to be excluded. When Judge Brian Gain decided that there was sufficient precedent for the technique and admitted the evidence, the defense was floored. Dimmock tried his best, reminding the jury that everything hinged on this contentious fingerprint testimony. It was the only thing that placed Hayden at the crime scene. "There's no other evidence to suggest Eric Hayden had ever been there,"[4] said Dimmock.

Senior Deputy Prosecutor James Konat was scathing in response. He told the jury, "You can enhance the fingerprints of anyone who is drunk enough or stupid enough to leave their fingerprints in the victim's blood."[5] The method used to enhance the print, he said, could have been duplicated with computer programs available in any software store.

The outcome was never in doubt. On January 10, 1996, after just two hours of deliberation, the jury convicted Hayden of murder. He was subsequently sentenced to 20 to 26 years in prison.

Speaking later, Berg expressed his satisfaction over Hayden's conviction. "I really believe that when he left there, he thought he'd gotten away with it."[6] And he defended his use of a technique that has revolutionized latent print identification, insisting that he merely found what was already there. "Now, did I find it with a magnifying glass? Did I find it with a brush and powder? No. I found it with a computer."[7]

Since this landmark case, other states have ruled that digital enhancement is an acceptable form of evidence, which has sparked a rush from crime labs around the country for similar software packages.

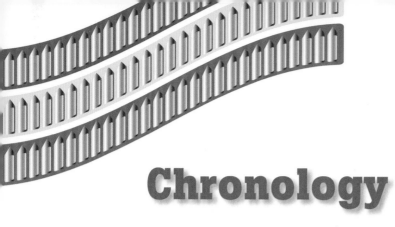

Chronology

1910   Edmond Locard establishes the world's first dedicated crime laboratory in Lyon, France; Albert S. Osborn writes a text entitled *Questioned Documents*, which leads to the acceptance of the scientific examination of documents in courts

1913   Victor Balthazard, professor of forensic medicine at the Sorbonne, publishes the first article on individualizing bullet markings

1915   Leone Lattes, professor at the Institute of Forensic Medicine in Turin Italy, develops the first antibody test for ABO blood groups; International Association for Criminal Identification, (to become The International Association of Identification (IAI), is organized in Oakland, California; Dr. Leone Lattes develops a procedure for determining the blood group of a dried bloodstain

1918   The Office of the Chief Medical Examiner opens in New York

1921   John Larson and Leonard Keeler design the portable polygraph

1924   August Vollmer, chief of police in Los Angeles, California, implements the first U.S. police crime laboratory

1925   The Bureau of Forensic Ballistics (BFB) is established in New York by Calvin Goddard, Charles Waite, Philip Gravelle, and John Fisher; the comparison microscope, an aid to firearms identification, is invented

1929	Calvin Goddard leaves the BFB to found the Scientific Crime Detection Laboratory on the campus of Northwestern University, Evanston, Illinois
1930	National fingerprint file is established in the United States by the Bureau of Investigation
1932	The Bureau of Investigation crime laboratory is created
1935	The Bureau of Investigation inaugurates the National Police Academy and becomes the Federal Bureau of Investigation (FBI)
1941	Murray Hill, of Bell Labs, initiates the study voice-print identification, a technique later refined by L. G. Kersta
1953	Dr. Paul L. Kirk publishes *Crime Investigation: Physical Evidence And The Police Laboratory,* a comprehensive reference work on all aspects of forensic science
1958	Neutron activation analysis is used for the first time in a homicide case
1960	First laser design to identify fingerprints (and other applications) is developed by U.S. physicist Theodore Maiman
1965	The FBI starts to develop the National Crime Information Center (NCIC) to collect data on wanted persons, stolen vehicles and stolen property
1967	NCIC is established
1971	Photo-fit system enabling witnesses to piece together facial features is developed by photographer Jacques Perry
1974	The detection of gunshot residue (GSR) using scanning electron microscopy with electron dispersive X-rays (SEM-EDX) technology is developed by J. E. Wessel, P. F. Jones, Q. Y. Kwan, R. S. Nesbitt and E. J. Rattin at Aerospace Corporation
1977	Fuseo Matsumur, a trace evidence examiner at the Saga Prefectural Crime Laboratory of the National

Police Agency of Japan, notices his own fingerprints developing on microscope slides while mounting hairs from a taxi driver murder case; he relates the information to coworker Masato Soba, a latent print examiner; Soba would later that year be the first to develop latent prints intentionally by "Superglue" fuming

1978 Electrostatic Detection Apparatus (ESDA) to expose handwriting impressions is developed by Bob Freeman and Doug Foster

1983 First use of personal computers in U.S. police patrol cars to provide quick information from National Crime Information Center

1984 Sir Alec Jeffreys develops the first DNA profiling test; it involves detection of a multilocus RFLP pattern

1986 In the first use of DNA to solve a crime, Jeffreys uses DNA profiling to identify Colin Pitchfork as the murderer of two young girls in the English Midlands; in the course of the investigation, DNA is first used to exonerate an innocent suspect

1987 DNA profiling introduced for the first time in a U.S. criminal court; Tommy Lee Andrews is convicted of a series of sexual assaults in Orlando, Florida

1991 Walsh Automation Inc., in Montreal, launches development of an automated imaging system called the Integrated Ballistics Identification System, or IBIS, for comparison of the marks left on fired bullets, cartridge cases, and shell casings; this system is subsequently developed for the U.S. market in collaboration with the Bureau of Alcohol, Tobacco, Firearms, and Explosives (ATF)

1993 In *Daubert et al. v. Merrell Dow Pharmaceuticals*, a U.S. federal court relaxes the *Frye* standard for admission of scientific evidence and confers on the judge a "gatekeeping" role

1998 An FBI DNA database, NIDIS, enabling interstate cooperation in linking crimes, is put into practice

1999 FBI upgrades its computerized fingerprint database and implements the Integrated Automated Fingerprint Identification System (IAFIS), allowing paperless submission, storage, and search capabilities directly to the national database maintained at the FBI

2007 The Office of the Chief Examiner in New York opens its high-sensitivity Forensic Biology DNA laboratory

Endnotes

Introduction
1. "Crime in the United States," Federal Bureau of Investigation, http://www.fbi.gov/ucr/cius2007/about/crime_clock.html (Accessed 5/18/2009).

Chapter 1
1. Brian Marriner, *Forensic Clues to Murder* (London: Arrow 1991), 224.

Chapter 2
1. Trial Transcript, para 514, l.10

Chapter 3
1. "Dr. Norris, 67, Dies of Sudden Illness," *New York Times*, September 12, 1935.
2. "Dr. Norris," *New York Times*, September 13, 1935.
3. "Find Slain Woman in Bronx Ash Pit: Missing 8 Months," *New York Times*, November 30, 1922.
4. "Says Becker Bared Pit Murder Plot," *New York Times*, December 20, 1922.
5. "Find Slain Woman in Bronx Ash Pit: Missing 8 Months."
6. Michael Baden and Judith Adler Hennessee, *Unnatural Death* (London: Sphere, 1991), 103–104.
7. "Norkin Re-Enacts Becker Murder," *New York Times*, December 4, 1922.
8. Ibid.
9. "Accused Each Other of Becker Murder," *New York Times*, December 1, 1922.
10. Ibid.

Chapter 5
1. "1440 Map Depicts the New World, *New York Times*, October 11, 1965, 1.
2. "1440 Vinland Map of America Is Declared Fake by Russian," *Washington Post*, April 16, 1966, A9.
3. Tom Zito, "Yale Says Vinland Map May Be Fake," *Washington Post*, January 26, 1974, A1.
4. Joe Arak, Viking Map Debate: Real or Fake?, http://www.cbsnews.com/stories/2002/07/30/tech/main516827.shtml (Accessed April 29, 2009).
5. Diane Scarponi, "Two Studies Battle Over Authenticity of Yale's Vinland Map," AP, July 30, 2002.
6. Diane Scarponi, "New Study Says Yale University's Vinland Map Is a Forgery," AP, July 29, 2002.

Chapter 6
1. Bill Smith, "Not Guilty! How the System Failed Patricia Stallings," *St. Louis Post Dispatch*, October 20, 1991.

2. Don W. Weber and Charles Bosworth Jr., *Precious Victims* (New York: Signet, 1991), 172.
3. "The Birth of the FBI's Technical Lab," Federal Bureau of Investigation, http://www.fbi.gov/libref/historic/history/birthtechlab.htm (Accessed May 8 15, 2009).
4. Ibid., 290.
5. Ibid.
6. No. 5-99-0250, Appellate Court Of Illinois, Fifth District.

Chapter 7
1. Bill Bass and Jon Jefferson, *Death's Acre: Inside the Legendary "Body Farm* (New York: Putnam's, 2003), 68.
2. Tara Young, "Seeking Ties that Bind," *Times-Picayune*, June 28, 1999.

Chapter 8
1. Stephen G. Badger, "Multiple Fire Deaths for 2005," *NFPA Journal*, http://www.nfpa.org/journalDetail.asp?categoryID=1255&itemID=29867&src=NFPAJournal (Accessed April 16, 2009).
2. John R. Hall Jr., "Intentional Fires and Arson," NFPA, http://www.nfpa.org/assets/files/PDF/ArsonSummary.pdf (Accessed April 16, 2009).
3. Jeffrey Sparshott, "Beltsville, Md., Fire Researcher Torches Items to Aid Investigators," *Tribune Business News*, September 24, 2004.
4. Ibid.
5. Jon Schmitz, "Witness Relates Arson Admission," *Pittsburgh Post-Gazette*, February 11, 1997.
6. Ibid.
7. Ann Belser, "Expert Links Arson to Firefighter's Death," *Pittsburgh-Post Gazette*, February 7, 1997.

Chapter 9
1. Susan Byrnes, "Family, Friends Mourn Kingsgate Slaying of Devoted Young Woman," *Seattle Post-Intelligencer*, May 16, 1995.
2. Susan Byrnes, "High Tech Corners Kirkland Murder Suspect," *Seattle Times*, June 2, 1995.
3. A. L. McRoberts, "Digital Image Processing as a Means of Enhancing Latent Fingerprints," Symposium on Latent Prints, FBI, July 7-10, 1987.
4. Kevin Ebi, "Neighbor Is Found Guilty," *Seattle Times*, January 10, 1996.
5. Ibid.
6. Jim Stewart, "The Hidden Clue," *60 Minutes*, http://www.foray.com/images/pdfs/60MinutesIIFinal-1.pdf (Accessed April 23, 2009).
7. Ibid.

Bibliography

Baden, Michael, and Judith Adler Hennessee. *Unnatural Death: Confessions of a Medical Examiner.* London: Sphere, 1991.
Baden, Michael, and Marion Roach. *Dead Reckoning: The New Science of Catching Killers.* New York: Simon & Schuster, 2001.
Cuthbert, C. R. M. *Science and the Detection of Crime.* London: Hutchinson, 1958.
Di Maio, Vincent J. M. *Gunshot Wounds: Practical Aspects of Firearms, Ballistics, and Forensic Techniques.* New York: Elsevier Science Publishing Co., 1985.
Dower, Alan. *Crime Scientist.* London: John Long, 1965.
Evans, Colin. *Blood on the Table: The Greatest Cases of New York City's Office of the Chief Medical Examiner.* New York: Berkley, 2008.
Evans, Colin. *The Casebook of Forensic Detection: How Science Solved 100 of the World's Most Baffling Crimes.* New York: Berkley, 2007.
Gaute, J. H. H., and Robin Odell. *Murder 'Whatdunit'.* London: Harrap, 1982.
Gaute, J. H. H., and Robin Odell. *The New Murderers' Who's Who.* New York: Dorset Press, 1979.
Gerber, Samuel, ed. *Chemistry and Crime: From Sherlock Holmes to Today's Courtroom.* Washington, D.C.: American Chemical Society, 1983.
Helpern, Milton, and Bernard Knight. *Autopsy.* London: Harrap, 1979.
Houts, Marshall. *Where Death Delights: The Story of Dr. Milton Helpern and Forensic Medicine.* New York: Coward-McCann, 1967.
Innes, Brian. *Bodies of Evidence.* Leicester, U.K.: Silverdale, 2000.
Knappman, Edward W., ed. *Great American Trials.* Detroit: Visible Ink, 1994.
Lane, Brian. *Encyclopedia of Forensic Science.* London: Headline, 1992.
Lee, Henry C., Timothy Palmbach, and Marilyn T. Miller. *Henry Lee's Crime Scene Handbook.* London: Academic, 2001.
Lewis, Alfred Allan. *The Evidence Never Lies: The Casebook of a Modern Sherlock Holmes.* New York: Holt, Rinehart and Winston, 1984.
Lustgarten, Edgar. *Verdict in Dispute.* London: Wingate, 1949.
Mackay, James. *Allan Pinkerton: The Eye Who Never Slept.* Edinburgh: Mainstream, 1996.
Marriner, Brian. *Forensic Clues to Murder.* London: Arrow, 1991.
Marten, Manuel Edward. *The Doctor Looks at Murder.* Garden City, N.Y.: Doubleday, 1937.
Morland, Nigel. *Science in Crime Detection.* London: Hale, 1958.

Noguchi, Thomas T., and Joseph DiMona. *Coroner.* Boston: G. K. Hall, 1984.
Noguchi, Thomas T., and Joseph DiMona. *Coroner At Large.* New York: Pocket Books, 1986.
Odell, Robin. *Science Against Crime.* London: Marshall Cavendish, 1982.
Saferstein, Richard. *Criminalistics: An Introduction to Forensic Science.* Upper Saddle River, N.J.: Prentice Hall, 1998.
Smyth, Frank. *Cause of Death: History of Murder Under the Microscope.* London: Pan Books, 1982.
Thompson, John. *Crime Scientist.* London: Harrap & Co., 1980.
Thorwald, Jürgen. *Dead Men Tell Tales.* London: Pan Books, 1968.
Thorwald, Jürgen. *The Century of the Detective.* New York: Harcourt, Brace & World, 1965.
Wecht, Cyril, Mark Curriden, and Benjamin Wecht. *Cause of Death.* New York: Dutton, 1993.
Wilson, Colin. *Written in Blood: A History of Forensic Detection.* New York: Carroll & Graf, 2003.
Wilson, Colin, and Patricia Pitman. *The Encyclopedia of Murder.* New York: Putnam's, 1962.
Wilson, Colin, and Donald Seaman. *The Encyclopedia of Modern Murder.* New York: Putnam's, 1983.

Further Resources

Print

Bass, Bill, and Jon Jefferson. *Death's Acre: Inside the Legendary Forensic Lab "Body Farm" Where the Dead Do Tell Tales.* New York: Putnam's, 2003. Records the origins and operation of Tennessee's famous "Body Farm."

Beavan, Colin. *Fingerprints: The Origins of Crime Detection.* New York: Hyperion, 2001. Describes the titanic battle to get fingerprints accepted as reliable evidence.

Borchard, Edwin M. *Convicting the Innocent: Errors of Criminal Justice.* New Haven, Conn.: Yale University Press, 1932. Groundbreaking investigation into 65 alleged miscarriages of justice.

Bougard, Thomas J. *Arson Investigation.* Springfield, Ill.: Thomas, 2004. Explains step-by-step how to process an arson crime scene.

Deutsch, Yvonne, ed. *Science Against Crime.* London: Marshall Cavendish, 1982. Highlights various forensic disciplines through sample cases.

Ebi, Kevin. (Various titles.) *Seattle Times,* December 26, 1995–January 10, 1996. Local background coverage of the Dawn Fehring murder.

Evans, Colin. *Slaughter on a Snowy Morn.* London: Icon, 2010. The first full-length book about the historic Stielow case.

Icove, David J., and John D. DeHaan, *Forensic Fire Scene Reconstruction.* Upper Saddle River, N.J.: Prentice Hall, 2008. Thorough examination of how fire works and the traces it leaves.

Keisch, Bernard. *The Atomic Fingerprint.* Washington, D.C.: Atomic Energy Commission, 1972. Covers the multi-layered uses of neutron activation analysis.

Lane, Brian. *The Encyclopedia of Forensic Science.* London: Headline, 1993. Overview of all aspects of forensic science.

Lee, Henry C., and Gaensslen, R.E., eds. *Advances in Fingerprint Technology.* Boca Raton, Florida: CRC, 2001. Records the latest developments in fingerprint technology.

Maples, William R., and Michael Browning. *Dead Men Do Tell Tales.* New York: Doubleday, 1994. A look at the day-to-day work of a forensic anthropologist.

People v. Stielow, 160, N.Y.S. 555-566 (July 1916). Court opinion in the Stielow case.

People v. Stielow, 161, N.Y.S. 599-616 (October 1916). Court opinion in the Stielow case.

"Professor Robert E. Jervis." *Journal of Radioanalytical and Nuclear Chemistry*, 179, 1 (1994). Provides background information about the career of Professor Jervis.

Sasser, Michael. *Fire Cops*. New York: Pocket Books, 1998. Case by case analysis of various arson incidents.

Seaver, Kirsten A. *Maps, Myths, and Men: The Story of the Vinland Map*. Palo Alto, Calif.: Stanford University Press, 2004. The author advances a controversial theory regarding the Vinland Map.

Skelton, R. A., Thomas E. Marston, and George D. Painter. *The Vinland Map and the Tartar Relation*. New Haven, Conn.: Yale University Press, 1996. A book arguing the authenticity of the Vinland Map.

Timbrell, John. *Introduction to Toxicology*. 3d ed. New York: Taylor & Francis, 2001. The science of detecting poisons and their effect on the human body.

Tully, Andrew W. *Inside the FBI*. New York: Dell, 1980. Describes the FBI's history and some of its most famous cases.

Washington v. Hayden, 38162-8-I (1998). Court opinion in the Hayden case.

Weber, Don W., and Charles Bosworth Jr. *Precious Victims*. New York: Signet, 1991. Full-length book about the Paula Sims case.

Online

Dowd, John T. "The Pillaged Grave of a Civil War Hero." Sons of Confederate Veterans. Available online. URL: http://www.tennessee-scv.org/shy.html. Accessed on May 6, 2009. In-depth account of the life of Colonel William M. Shy.

Hadingham, Evan. "The Viking Deception." Nova Science Programming on Air and Online. Available online. URL: http://www.pbs.org/wgbh/nova/vinland. Accessed on June 19, 2009. Comprehensive Web site dealing with the Vinland Map.

Illinois v. Sims, 750 N.E.2d 320, 323 (Ill. App. Ct. 2001). "People of the State of Illinois v. Paula J. Sims." Official Website of the State of Illinois. Available online. URL: http://www.state.il.us/court/opinions/appellatecourt/2001/5thdistrict/june/html/5990250.htm. Accessed on May 15, 2009. Court opinion in the Sims case.

Index

A
ABO blood typing system 13
absorption-elution test 15
accelerants 99, 100
accelerator mass spectrometer 66, 67
accumulated degree days (ADDs) 89
Alcohol, Tobacco, Firearms, and Explosives (ATF) 95
    Fire Research Laboratory 95–97
    mobile laboratory 105
    National Laboratory Center 95, 96
    Pittsburgh fire investigation 98–101, 104–105
Alcohol and Tobacco lab (National Laboratory Center, Maryland) 96
alcohol content, in wet blood 15
alligatoring effect burns 99
Almodovar, Anibal 90, 91
Almodovar, Louise 90, 91
American Journal of Police Science 35–36
amido black dye treatment 109, 115, 116
anatase 65–66, 69
Anglin, Arthur L. 56
animal blood 14, 24
Anthropological Research Facility (University of Tennessee, Knoxville). *See* Body Farm
anthropometry 22–23
anti-forensics programs 112
antigens 14
antiserum tests 14
Appel, Charles 78, 79
arsenic 17–21
"arsenic mirror" 18, 21

arson 94–105
    motives for 94
    Pittsburgh fire investigation 98–101, 104–105
    Tetzner case 102–103
art forgeries 60–61, 63
ATF. *See* Alcohol, Tobacco, Firearms, and Explosives
ATF Forensic Science lab (National Laboratory Center, Maryland) 96
Attenburger, David 79–80
autopsies 38, 41

B
Baden, Michael 44
Bajo, Julie 114
ballistics 27–37
ballistics experts 32
Balthazard, Victor 28
Bass, William M. 82, 84–86, 88–93
Bausch & Lomb Company 36
Becker, Abraham 39, 42–47
Becker, Jennie 39, 42–47
Beinecke Rare Book and Manuscript Library (Yale University) 64
Bellevue Hospital 39, 41, 47–48
benzidine 14
Berg, Erik 109, 112–117
Berley, George 32
Berry, Whitney 88
Bertillon, Alphonse 20–23
Bertillon, Jacques 22
Bertillonage 20, 23
blood evidence 13–15, 24
blood spatter analysis 21
bloodstain pattern analysis 13, 15
blood types 13–14

blood typing 15
blowflies 89–93
bodily fluids 15
Bodle, George 19
Bodle, John 19
Body Farm (Anthropological Research Facility, University of Tennessee, Knoxville) 82–83, 85, 86
Bond, George 33, 36
Boston, Massachusetts 38
Bouchard, Gaetane 51–56
Bouchard, Wilfrid 51, 54–56
British Museum 64
Brooks, Thomas 97–98
Brown, Gregory 101, 104–105
Brown, Katherine 69
Buckner, Darlene 96, 99, 101, 104–105
Buckner, Ronald 96
Buckner family 96, 99, 101
bullet identification
    ballistics 27–38
    with NAA 58
Bunsen, Robert 52
Bunsen-Kirchoff Award 53
Bureau of Forensic Ballistics 34, 35
Bureau of Investigation, U.S. 25, 36, 78
Butts, Harry F. 33, 36

C

Cahill, Thomas 68
carbon isotopes 66, 67
Carter, Howard 60
Case, Mary 74, 75
CBSU (Chemical and Biological Sciences Unit) 79
Celldar system 57
cell phones 57
Chemical and Biological Sciences Unit (CBSU) 79
Chicago, Illinois 38
chromatography 9, 100
Clark, Robin 69
CODIS database 79
color tests, for bloodstains 14
comparison microscope 32, 34–35
computer forensics
    crime scene video reconstruction 114, 115

data gathering in 110–112
    Fehring murder case 109, 112–117
cone calorimeter 95
Conley, Glen, Jr. 88
Connor, Harold 87
Conroy, Patricia 97–98
Copeland, Joseph J. 90, 91
coroners 38–39
corruption, in coroner system 38–39
Counterterrorism and Forensic Science Research Unit (FBI Laboratory) 79
crimes, in the United States 11, 12
crime scenes 9
crime scene video reconstruction 114, 115
Cryptanalysis Unit (FBI Laboratory) 79
crystalline tests, for bloodstains 14
"CSI Effect" 7
CTIA (International Association for the Wireless Telecommunications Industry) 57
cybercriminals 111

D

Daubert et al. v. Merrell Dow Pharmaceuticals 106
decomposition of bodies 83, 86, 89
de Medici, Catherine 17
dental records 21
dermal nitrate test 58
die lines 80
digital enhancement 109, 112–117
digital photography 116
Dimmock, Andrew 114, 116–117
direct blood typing 15
DNA analysis/testing 9–10
    CODIS database 79
    at Forensic Biology DNA lab 46
DNA typing 9, 15
documents, authenticity of 62–70
Dooley, Mick 74
Dreyfus, Alfred 23
drug content, in wet blood 15

E

electron scanning microscope 9
electrostatic detection apparatus (ESDA) 9, 62
Elias, Anna 39, 42

130 CRIME LAB

Eriksson, Leif 64
ESDA (electrostatic detection apparatus) 9, 62
evidence 10, 11, 71–81. *See also* science-based evidence
evidence interpretation 11
evidential value 9, 11, 15
expert witnesses
    in 19th century Europe 7
    Albert Hamilton 30–31, 33, 36

F

Faraday, Michael 19
Faurot, James A. 33
FBI Laboratory 79
Federal Bureau of Investigation (FBI)
    chemical profile standards 58
    crime lab 25–26, 78, 79
    and LCN analysis 10
    and Sims case 77, 80
Fehring, Dawn 107, 108
Ferrajoli de Ry, Enzo 64, 70
fingerprint identification 9, 23, 78
    digital enhancement 109, 112–117
    reliability of 21
firearms identification 27, 28, 34, 35
    ballistics 27–38
    by grips 32
    by "gun experts" 30, 31, 33
Fire Research Laboratory (FRL) 95–97
fires 94–97. *See also* arson
Fischer, Joseph 69–70
Fisher, John H. 34
"Forensic Ballistics" (Goddard) 34
forensic ballistics (term) 34
Forensic Biology DNA lab (New York City OCME) 46
forensic science 9, 10, 21, 26
forgeries 60–70
Fournier, Michael 87
FRL (Fire Research Laboratory) 95–97
Frye hearing 115
Frye v. U.S. 106
fusors 58
future of crime labs 46

G

Gain, Brian 116
garbage bags 76–80

Garçon à la Pipe (Picasso) 60, 61
gasoline, as accelerant 99
gel tests 14
Gettler, Alexander O. 44, 47, 90
Goddard, Calvin H. 35, 78
governments 9
grasses 90, 91
Gravelle, Philip O. 34–35
grave robbing 84
Green, Nelson 29–31, 35, 37
grips, identifying weapon by 32
grooves (rifling) 27
gunshot residue tests 58
gunshot wounds 21

H

Haag, Lucien 115
half-life 66, 67
Hall, Albert Llewellyn 28
Hall-Mills murder case 35
Hamilton, Albert 30–31, 33, 36, 37
Hayden, Eric 108–109, 112–117
heat-release rate test 95
Heinrich, Edward Oscar 25
helixometer 35
Hevesy, George de 49, 50
history
    of crime labs 17–26
    of science-based evidence 7–9
Hofmann, Eduard von 40
Hollien, Harry 114
Holshue, Daniel 108–109, 113, 115, 116
Hoover, J. Edgar 36, 78
human blood 15, 24
human body
    fire-resistance of 94
    postmortem changes in 21, 82–93
Humanitarian Cult 31

I

identifying samples 11
image enhancement 109, 116. *See also* digital enhancement
image restoration 116
indirect blood typing 15
"inheritance powder" 17
inquest juries 38
insect depredation 89–93

Institute of Forensic Medicine (University of Leipzig) 102
insurance
    arson for 94, 99, 102
    murder for 86–93
International Association for the Wireless Telecommunications Industry (CTIA) 57
Iowa Department of Criminal Investigation 37
isotope sources 58

J

Jervis, Robert E. 55
Jones, William A. 33
Judicial Identification Service 23
junk science 11, 72, 73
juries 7, 30, 38

K

Kennard, Karl S. 43–45
Kennedy, John F. 58
Keystone Insurance Company 99
King, Erwin 31, 33, 37
Kirchoff, Gustav 52, 53
Kockel, Richard 102–103
Koenigstein, François Claudius ("Ravachol") 23
Kolenda, Marc 97–98
Kolisko, Alexander 40
Konat, James 117

L

Laboratoire Intérregional de Police Technique (Lyon) 24
Lacassagne, Alexandre 21, 24
Lafarge, Marie 19
lands (rifling) 27
Landsteiner, Karl 13–14, 24
La Santé Prison 22
LCN (low copy number analysis) 9–10
Levi, Hilde 49, 50
Libby, Willard F. 66
lie detectors 106
light microscope 92
Locard, Edmond 24, 25
Locard's Exchange Principle 24, 25
Los Angeles County Sheriff's Department 116

low copy number analysis (LCN) 9–10
luminol 14, 15

M

manslaughter 114
Marsh, James 18–21
Maximilian I (Holy Roman emperor) 27
Maybrick, Florence 7
McClung, Soraya 12
McCrone, Walter 65, 66, 68, 69
medical examiner's office (New York City) 39
medical jurisprudence 14
medical testimony 8
Mellon, Paul 64
methylmalonic acidemia (MMA) 72
Metzger, Johann 17–18
Miller, Spencer 31
Mitchell, Artie 113, 115
Mitchell, Jim 113, 115
Mitchell, John Purroy 38–39
mitochondrial DNA extraction 77
MMA (methylmalonic acidemia) 72
Morton, Charles 115
mug shots 22–23
multidisciplinary crime labs 35–36
murder
    Almodovar case 90, 91
    in arson cases 101, 103
    Fehring case 107–109, 112–117
    manslaughter vs. 114
    Perry case 86–93
    Sims case 71–81
    Stallings case 72, 73
    in the United States 11
Mutual of New York 88

N

NAA (neutron activation analysis) 9, 49–59
Napoleon Bonaparte (emperor of the French) 45
National Ballistics Intelligence Service 28
National Fire Protection Association 94
National Laboratory Center (Maryland) 95, 96

neutron activation analysis (NAA) 9, 49–59
Nevsky, Vladimir 64–64
New York City 38–39, 41, 44–48
New York Life Insurance 87
New York Police Department (NYPD) 33
New York Times 41
Nichols, David 79
Norkin, Reuben 42–43, 47
Norris, Charles 39–41, 44–46
NYPD (New York Police Department) 33

O

O'Connell, Clarence 31, 33, 37
Office of the Chief Medical Examiner (OCME) 39, 41, 44–48
Olin, Jacqueline 68
orthotolodine 14

P

pace of development 9
Parkman, George 21
pathology 40
percussion cap, for naval guns 19
Perry, Annie 86–93
Perry, Darrell 86–93
Perry, Krystal 86, 87, 93
Petraitis, William J. 105
Phelps, Charles 29
phenolphthalein 14
physical contact, traces left by 24
Picasso, Pablo 60, 61
Pichette, J. A. 56
Pittsburgh Fire Department 96–98
poisonings 17–18, 72, 73
polygraphs 106
Poly Tech 77
Poser, Max 36
postmortem changes in human body 21
    decomposition 83, 86
    insect depredation 89–93
    and time of death 82–93
Précis de Médicine Légale 21
private crime lab facilities 26
Prochaska, Alice 70
professional poisoners 17
provenance 60, 67
pyrolysis gas chromatograph 100

R

radiation, in NAA 55
radio carbon dating 66, 67
radio footprints 57
Raman microscope 69
"Ravachol" 23
Rees, Melvin David 32
"referee method" 49
rifling 27–28
Rinaldo, Piero 73
Robillard, Allen 80
Rose, Valentine 18
Royal Armaments Research and Development Establishment (Fort Halstead, Kent, England) 9
royal families of Europe 17
Royal Military Academy (Woolwich, Great Britain) 19
Rubenstein, Alan Michael 86–88, 91–93
Rubenstein, Doris 86, 87
Rubenstein, Tonya 91

S

Sacco, Nicola 35
saliva 15
Saunders, Charles 74
Schwartz, Charles, Jr. 88
science-based crime detection 21, 22
science-based evidence 7
    admissibility of 106, 111, 114–117
    history of 7–9
Scientific Crime Detection Laboratory (Northwestern University) 35–36
scintillation counter 55
Seaver, Kirsten 69–70
semen 15
sequence of writing, proving 62
Sercone, David S. 104–105
serologists 14–15
sexual assaults, in the United States 11
Shoemaker, James 72, 73
Shroud of Turin 68
Shy, William 84, 85
Sims, Heather 72–74, 76, 80
Sims, Loralei 75, 81
Sims, Paula 71–81

Sims, Randy 74, 76
Sims, Robert 71–76
smoke, study of 96, 97
Sokolsky, Sidney 105
specialized crime labs 9, 26, 110
spectrophotometer 52
spectroscopy 52, 53
Speculum Historale 63, 64
Stallings, David 72, 73
Stallings, Patricia 72
Stallings, Ryan 72, 73
stand-alone crime investigation facilities 24, 25
Stella Matutina (Feldkirch, Austria) 70
Stielow, Charles 29–31, 33, 35–37
stomach contents 18, 44, 47
striations (ballistics) 28
Supreme Court, U.S. 106
Sweeney, Shaun E. 105

T

Tacoma Police Department 109
Tartar Relation 63, 64
Technical Crime Laboratory (FBI) 25–26, 36, 78, 79
Technical Laboratory (United States Bureau of Investigation) 25
temperature-decomposition ratio 89
test benches 9
Tetzner, Erich 102, 103
3D reconstruction of crime scenes 114, 115
time since death (TSD) 82–93
toxicology kit 45
toxicology tests 9, 17–21
trace evidence 53
TSD (time since death) 82–93
Turin Shroud 68

U

Uhlenhuth, Paul 24
University College (London) 69
University of Leon 21

V

van Meegeren, Hans 61, 63
Vanzetti, Bartolomeo 35
video reconstruction of crime scenes 114, 115
Video Spectral Comparator 5000 (VSC5000) 62
Vinland Map 63–70
violent crimes, in the United States 11, 12
Virchow, Rudolf 40
Vollman, John 51, 54–56
VSC 5000 (Video Spectral Comparator 5000) 62

W

Waite, Charles E. 33, 35
Wallstein, Leonard M. 39
Ward, Emily 93
Warner, Rex 77
Washington State 106–107
Webster, Joseph 21
West Virginia State Police Crime Lab 12
wet blood, evidentiary value of 15
White, David A. 30
Whitman, Charles 31, 33
Witten, Laurence C. 64
Wolcott, Margaret 29
Wright, Keith 101

Y

Yale University 64, 65, 68

About the Author

Colin Evans is a writer specializing in criminal investigations and forensics. He has written numerous articles and books, including *Blood on the Table: The Greatest Cases of New York City's Office of the Chief Medical Examiner*, *The Casebook of Forensic Detection*, and for Chelsea House, *Crime Scene Investigation*, *Evidence*, and *Trials and the Courts*. He has been a major contributor to *Courtroom Drama; Great World Trials;* and *Great American Trials*. Evans lives in the United Kingdom.